‖‖‖ ‖ ‖‖‖‖‖‖‖‖‖‖‖‖ ‖‖‖ ‖‖‖‖ ‖‖
I0005730

About the Author

Jiashi Yang is a Full Professor at the Department of Mechanical and Materials Engineering of University of Nebraska–Lincoln. He received his bachelor's and master's degrees from Tsinghua University in 1982 and 1985, and his Ph.D. from Princeton University in 1994. Then he did postdoctoral research at University of Missouri-Rolla and Rensselaer Polytechnic Institute. He worked as an engineer with Motorola, Inc. prior to joining University of Nebraska–Lincoln in 1997. His research area is the mechanics of electro-magneto-mechanical structures and devices. His previous books include *Mechanics of Piezoelectric Structures*, *Mechanics of Functional Materials* and *Theory of Electromagnetoelasticity* with World Scientific, as well as *An Introduction to the Theory of Piezoelectricity* and *Analysis of Piezoelectric Semiconductor Structures* with Springer.

Contents

Chapter 1

ELASTICITY

In this chapter, we develop the nonlinear theory of elasticity for large deformation. It is a special case of the nonlinear theories of magnetoelasticity and ferromagnetoelasticity in Chapters 3, 5 and 10, and serves as a preparation for Chapter 3. The two-point Cartesian tensor notation is used along with the summation convention for repeated tensor indices and the convention that a comma followed by an index denotes partial differentiation with respect to the coordinate associated with the index [1]. The chapter is not meant to be a complete treatment of the theory of elasticity. Only those results that are needed for the rest of the book are presented.

1.1 Kinematics

Consider a deformable continuum which, in the reference configuration at time t_0, occupies a region V with a boundary surface S (see Fig. 1.1). $\mathbf{N}$ is the outward unit normal of S. The position of a material point in this state is denoted by a position vector $\mathbf{X} = X_K \mathbf{I}_K$ in a rectangular coordinate system. X_K denote the reference or material coordinates of the material point. They are a continuous labeling of material particles so that they are identifiable. At time t, the body occupies a region v with a boundary surface s and an outward unit normal $\mathbf{n}$. The current position of the material point associated with $\mathbf{X}$ is given by $\mathbf{y} = y_k \mathbf{i}_k$ which denotes the present or spatial coordinates of the material point.

The coordinate systems are assumed to be othonormal, i.e.,

$$\mathbf{i}_k \cdot \mathbf{i}_l = \delta_{kl}, \quad \mathbf{I}_K \cdot \mathbf{I}_L = \delta_{KL}, \tag{1.1.1}$$

where δ_{kl} and δ_{KL} are the Kronecker delta. In matrix notation,

$$[\delta_{kl}] = [\delta_{KL}] = \begin{bmatrix} 1 & 0 & 0 \\ 0 & 1 & 0 \\ 0 & 0 & 1 \end{bmatrix}. \tag{1.1.2}$$

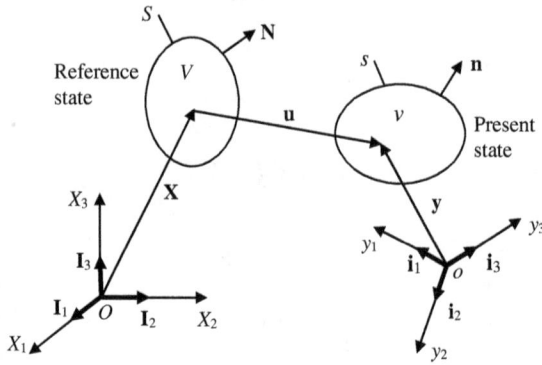

Fig. 1.1. Motion of a continuum and coordinate systems.

The transformation coefficients between the unit vectors of the two coordinate systems are denoted by

$$\mathbf{i}_k \cdot \mathbf{I}_L = \delta_{kL} = \delta_{Lk} = \mathbf{I}_L \cdot \mathbf{i}_k. \tag{1.1.3}$$

In the rest of this book, the two coordinate systems are chosen to be coincident, i.e.,

$$o = O, \quad \mathbf{i}_1 = \mathbf{I}_1, \quad \mathbf{i}_2 = \mathbf{I}_2, \quad \mathbf{i}_3 = \mathbf{I}_3. \tag{1.1.4}$$

Then δ_{kL} becomes the Kronecker delta. A vector can be resolved into rectangular components in different coordinate systems. For example, we can also write

$$\mathbf{y} = y_K \mathbf{I}_K, \tag{1.1.5}$$

with

$$y_M = \delta_{Mi} y_i. \tag{1.1.6}$$

The motion of the body is described by

$$y_i = y_i(\mathbf{X}, t). \tag{1.1.7}$$

The displacement vector $\mathbf{u}$ of a material point is defined by

$$\mathbf{y} = \mathbf{X} + \mathbf{u}, \tag{1.1.8}$$

or

$$y_i = \delta_{iM}(X_M + u_M). \tag{1.1.9}$$

An infinitesimal material line element $\mathbf{dX}$ at t_0 deforms into the following line element at t:

$$dy_i|_{t \text{ fixed}} = y_{i,K} dX_K, \tag{1.1.10}$$

where the deformation gradient,

$$y_{k,K} = \delta_{kK} + u_{k,K}, \tag{1.1.11}$$

is a two-point tensor depending on two coordinate systems. The following determinant is called the Jacobian of the deformation:

$$J = \det(y_{k,K}) = \varepsilon_{ijk} y_{i,1} y_{j,2} y_{k,3}$$

$$= \varepsilon_{KLM} y_{1,K} y_{2,L} y_{3,M} = \frac{1}{3!} \varepsilon_{klm} \varepsilon_{KLM} y_{k,K} y_{l,L} y_{m,M} > 0, \tag{1.1.12}$$

where ε_{klm} and ε_{KLM} are the permutation symbol defined by

$$\varepsilon_{ijk} = \mathbf{i}_i \cdot (\mathbf{i}_j \times \mathbf{i}_k)$$

$$= \begin{cases} 1 & i,j,k = 1,2,3; \quad 2,3,1; \quad 3,1,2, \\ -1 & i,j,k = 3,2,1; \quad 2,1,3; \quad 1,3,2, \\ 0 & \text{otherwise.} \end{cases} \tag{1.1.13}$$

The following relationship exists ($\varepsilon - \delta$ identity) [2]:

$$\varepsilon_{ijk} \varepsilon_{pqr} = \begin{vmatrix} \delta_{ip} & \delta_{iq} & \delta_{ir} \\ \delta_{jp} & \delta_{jq} & \delta_{jr} \\ \delta_{kp} & \delta_{kq} & \delta_{kr} \end{vmatrix}. \tag{1.1.14}$$

As a special case, when $i = p$, Eq. (1.1.14) reduces to

$$\varepsilon_{ijk} \varepsilon_{iqr} = \delta_{jq} \delta_{kr} - \delta_{jr} \delta_{kq}. \tag{1.1.15}$$

It can be shown that [3]

$$J = \frac{1}{6} \left[2 \frac{\partial y_K}{\partial X_L} \frac{\partial y_L}{\partial X_M} \frac{\partial y_M}{\partial X_K} - 3 \frac{\partial y_K}{\partial X_K} \frac{\partial y_L}{\partial X_M} \frac{\partial y_M}{\partial X_L} + \left(\frac{\partial y_M}{\partial X_M} \right)^3 \right]. \tag{1.1.16}$$

It can be verified directly that for all L, M and N the following is true:

$$\varepsilon_{ijk} y_{i,L} y_{j,M} y_{k,N} = J \varepsilon_{LMN}. \tag{1.1.17}$$

From Eq. (1.1.17), the following can be shown:

$$\varepsilon_{ijk} y_{j,M} y_{k,N} = J \varepsilon_{LMN} X_{L,i},$$

$$\varepsilon_{ijk} y_{k,N} = J \varepsilon_{LMN} X_{L,i} X_{M,j}. \tag{1.1.18}$$

Proof. Multiplying both sides of Eq. (1.1.17) by $X_{L,r}$, we have

$$\varepsilon_{ijk} y_{i,L} X_{L,r} y_{j,M} y_{k,N} = J \varepsilon_{LMN} X_{L,r}. \tag{1.1.19}$$

Then

$$\varepsilon_{ijk} \delta_{ir} y_{j,M} y_{k,N} = J \varepsilon_{LMN} X_{L,r}, \tag{1.1.20}$$

$$\varepsilon_{rjk} y_{j,M} y_{k,N} = J \varepsilon_{LMN} X_{L,r}. \tag{1.1.21}$$

Replacing the index r by i gives Eq. $(1.1.18)_1$. Similarly, Eq. $(1.1.18)_2$ can be proved by multiplying both sides of Eq. $(1.1.18)_1$ by $X_{M,p}$.

The following relationship can then be established:

$$\varepsilon_{ijk} \varepsilon_{LMN} y_{j,M} y_{k,N} = 2 J X_{L,i}. \tag{1.1.22}$$

Proof. The multiplication of both sides of Eq. $(1.1.18)_1$ by ε_{PMN} gives

$$\varepsilon_{ijk} \varepsilon_{PMN} y_{j,M} y_{k,N} = J \varepsilon_{LMN} \varepsilon_{PMN} X_{L,i} = J \varepsilon_{MNL} \varepsilon_{MNP} X_{L,i}$$

$$= J(\delta_{NN} \delta_{LP} - \delta_{NP} \delta_{LN}) X_{L,i} = J(3 \delta_{LP} - \delta_{LP}) X_{L,i}$$

$$= 2 J X_{P,i}, \tag{1.1.23}$$

where Eq. (1.1.15) has been used. Replacing the index P by L gives Eq. (1.1.22).

The derivative of the Jacobian with respect to one of its elements is

$$\frac{\partial J}{\partial y_{i,L}} = J X_{L,i}. \tag{1.1.24}$$

Proof. From Eq. (1.1.12),

$$6 \frac{\partial J}{\partial y_{p,Q}} = \varepsilon_{ijk} \varepsilon_{LMN} \delta_{ip} \delta_{LQ} y_{j,M} y_{k,N}$$

$$+ \varepsilon_{ijk} \varepsilon_{LMN} y_{i,L} \delta_{jp} \delta_{MQ} y_{k,N} + \varepsilon_{ijk} \varepsilon_{LMN} y_{i,L} y_{j,M} \delta_{kp} \delta_{NQ}$$

$$= \varepsilon_{pjk} \varepsilon_{QMN} y_{j,M} y_{k,N} + \varepsilon_{ipk} \varepsilon_{LQN} y_{i,L} y_{k,N} + \varepsilon_{ijp} \varepsilon_{LMQ} y_{i,L} y_{j,M}$$

$$= 3 \varepsilon_{pjk} \varepsilon_{QMN} y_{j,M} y_{k,N} = 3(2 J X_{Q,p}), \tag{1.1.25}$$

where Eq. (1.1.22) has been used.

With Eq. (1.1.22), it can also be shown that

$$(J X_{K,k})_{,K} = 0. \tag{1.1.26}$$

Proof. The differentiation of both sides of Eq. (1.1.22) with respect to X_L gives

$$2(JX_{L,i}),_L = \varepsilon_{ijk}\varepsilon_{LMN}(y_{j,ML}y_{k,N} + y_{j,M}y_{k,NL}) = 0, \qquad (1.1.27)$$

because

$$\varepsilon_{LMN}y_{j,ML} = 0, \quad \varepsilon_{LMN}y_{k,NL} = 0. \qquad (1.1.28)$$

Similarly, the following is true:

$$(J^{-1}y_{k,K}),_k = 0. \qquad (1.1.29)$$

The lengths of a material line element before and after deformation are given by

$$(dL)^2 = dX_K dX_K = \delta_{KL}dX_K dX_L, \qquad (1.1.30)$$

and

$$(dl)^2 = dy_i dy_i = y_{i,K}dX_K y_{i,L}dX_L$$
$$= C_{KL}dX_K dX_L, \qquad (1.1.31)$$

where C_{KL} is the deformation tensor defined by

$$C_{KL} = y_{k,K}y_{k,L} = C_{LK}. \qquad (1.1.32)$$

From Eqs. (1.1.30) and (1.1.31), we have

$$(dl)^2 - (dL)^2 = (C_{KL} - \delta_{KL})dX_K dX_L$$
$$= 2E_{KL}dX_K dX_L, \qquad (1.1.33)$$

where the finite strain tensor E_{KL} is defined by

$$E_{KL} = (C_{KL} - \delta_{KL})/2 = (y_{i,K}y_{i,L} - \delta_{KL})/2$$
$$= (u_{K,L} + u_{L,K} + u_{M,K}u_{M,L})/2 = E_{LK}. \qquad (1.1.34)$$

The unabbreviated form of Eq. (1.1.34) is

$$E_{11} = u_{1,1} + (u_{1,1}u_{1,1} + u_{2,1}u_{2,1} + u_{3,1}u_{3,1})/2,$$
$$E_{22} = u_{2,2} + (u_{1,2}u_{1,2} + u_{2,2}u_{2,2} + u_{3,2}u_{3,2})/2, \qquad (1.1.35)$$
$$E_{33} = u_{3,3} + (u_{1,3}u_{1,3} + u_{2,3}u_{2,3} + u_{3,3}u_{3,3})/2,$$
$$E_{23} = (u_{2,3} + u_{3,2} + u_{1,2}u_{1,3} + u_{2,2}u_{2,3} + u_{3,2}u_{3,3})/2,$$
$$E_{31} = (u_{3,1} + u_{1,3} + u_{1,3}u_{1,1} + u_{2,3}u_{2,1} + u_{3,3}u_{3,1})/2, \qquad (1.1.36)$$
$$E_{12} = (u_{1,2} + u_{2,1} + u_{1,1}u_{1,2} + u_{2,1}u_{2,2} + u_{3,1}u_{3,2})/2.$$

At the same material point, consider two material line elements $d\overset{(1)}{\mathbf{X}}$ and $d\overset{(2)}{\mathbf{X}}$ which deform into $d\overset{(1)}{\mathbf{y}}$ and $d\overset{(2)}{\mathbf{y}}$. The area of the parallelogram spanned by $d\overset{(1)}{\mathbf{X}}$ and $d\overset{(2)}{\mathbf{X}}$, and that by $d\overset{(1)}{\mathbf{y}}$ and $d\overset{(2)}{\mathbf{y}}$, can be represented by the following vectors, respectively,

$$N_L dS = dS_L = \varepsilon_{LMN} d\overset{(1)}{X}{}_M d\overset{(2)}{X}{}_N, \qquad (1.1.37)$$

$$n_i ds = ds_i = \varepsilon_{ijk} d\overset{(1)}{y}{}_j d\overset{(2)}{y}{}_k. \qquad (1.1.38)$$

They are related by

$$ds_i = J X_{L,i} dS_L. \qquad (1.1.39)$$

Proof.

$$ds_i = \varepsilon_{ijk} d\overset{(1)}{y}{}_j d\overset{(2)}{y}{}_k$$

$$= \varepsilon_{ijk} y_{j,M} d\overset{(1)}{X}{}_M y_{k,N} d\overset{(2)}{X}{}_N = \varepsilon_{ijk} y_{j,M} y_{k,N} d\overset{(1)}{X}{}_M d\overset{(2)}{X}{}_N$$

$$= J X_{L,i} \varepsilon_{LMN} d\overset{(1)}{X}{}_M d\overset{(2)}{X}{}_N = J X_{L,i} dS_L, \qquad (1.1.40)$$

where Eq. (1.1.18)$_1$ has been used.

At the same material point, consider three material line elements $d\overset{(1)}{\mathbf{X}}$, $d\overset{(2)}{\mathbf{X}}$ and $d\overset{(3)}{\mathbf{X}}$ which deform into $d\overset{(1)}{\mathbf{y}}$, $d\overset{(2)}{\mathbf{y}}$ and $d\overset{(3)}{\mathbf{y}}$. The volume of the parallelepiped spanned by $d\overset{(1)}{\mathbf{X}}$, $d\overset{(2)}{\mathbf{X}}$ and $d\overset{(3)}{\mathbf{X}}$, and that by $d\overset{(1)}{\mathbf{y}}$, $d\overset{(2)}{\mathbf{y}}$ and $d\overset{(3)}{\mathbf{y}}$, are related by

$$dv = J dV. \qquad (1.1.41)$$

Proof.

$$dv = d\overset{(1)}{\mathbf{y}} \cdot (d\overset{(2)}{\mathbf{y}} \times d\overset{(3)}{\mathbf{y}}) = \varepsilon_{ijk} d\overset{(1)}{y}{}_i d\overset{(2)}{y}{}_j d\overset{(3)}{y}{}_k$$

$$= \varepsilon_{ijk} y_{i,L} d\overset{(1)}{X}{}_L y_{j,M} d\overset{(2)}{X}{}_M y_{k,N} d\overset{(3)}{X}{}_N$$

$$= \varepsilon_{ijk} y_{i,L} y_{j,M} y_{k,N} d\overset{(1)}{X}{}_L d\overset{(2)}{X}{}_M d\overset{(3)}{X}{}_N$$

$$= J \varepsilon_{LMN} d\overset{(1)}{X}{}_L d\overset{(2)}{X}{}_M d\overset{(3)}{X}{}_N$$

$$= J d\overset{(1)}{\mathbf{X}} \cdot (d\overset{(2)}{\mathbf{X}} \times d\overset{(3)}{\mathbf{X}}) = J dV, \qquad (1.1.42)$$

where Eq. (1.1.17) has been used.

The velocity and acceleration of a material point are given by the following material time derivatives:

$$v_i = \frac{dy_i}{dt} = \dot{y}_i = \left.\frac{\partial y_i(\mathbf{X},t)}{\partial t}\right|_{\mathbf{X}\text{ fixed}},$$

$$a_i = \frac{dv_i}{dt} = \dot{v}_i = \ddot{y}_i = \left.\frac{\partial^2 y_i(\mathbf{X},t)}{\partial t^2}\right|_{\mathbf{X}\text{ fixed}}. \tag{1.1.43}$$

The deformation rate tensor d_{ij} and the spin tensor ω_{ij} are introduced by decomposing the velocity gradient into symmetric and antisymmetric parts, i.e.,

$$\partial_i v_j = v_{j,i} = d_{ij} + \omega_{ij},$$

$$d_{ij} = \frac{1}{2}(v_{j,i} + v_{i,j}), \quad \omega_{ij} = \frac{1}{2}(v_{j,i} - v_{i,j}). \tag{1.1.44}$$

We also have

$$\frac{d}{dt}(dy_i) = \frac{d}{dt}\left(\frac{\partial y_i}{\partial X_K} dX_K\right) = \frac{\partial}{\partial X_K}\left(\frac{dy_i}{dt}\right) dX_K$$

$$= \frac{\partial}{\partial X_K}(v_i) dX_K = v_{i,K} dX_K = v_{i,j} y_{j,K} dX_K. \tag{1.1.45}$$

The strain rate and the deformation rate are related by

$$\dot{E}_{KL} = d_{ij} y_{i,K} y_{j,L}. \tag{1.1.46}$$

Proof.

$$\dot{E}_{KL} = \frac{1}{2}(\dot{y}_{i,K} y_{i,L} + y_{i,K} \dot{y}_{i,L}) = \frac{1}{2}(v_{i,K} y_{i,L} + y_{i,K} v_{i,L})$$

$$= \frac{1}{2}(v_{i,j} y_{j,K} y_{i,L} + y_{i,K} v_{i,j} y_{j,L})$$

$$= \frac{1}{2}(v_{j,i} y_{i,K} y_{j,L} + y_{i,K} v_{i,j} y_{j,L}) \tag{1.1.47}$$

$$= \frac{1}{2}(v_{j,i} + v_{i,j}) y_{i,K} y_{j,L} = d_{ij} y_{i,K} y_{j,L}.$$

The material derivative of the Jacobian is given by

$$\dot{J} = J v_{k,k}. \tag{1.1.48}$$

Proof. From Eq. (1.1.12),

$$\dot{J} = \frac{1}{6}\varepsilon_{klm}\varepsilon_{KLM}\left(v_{k,K}y_{l,L}y_{m,M} + y_{k,K}v_{l,L}y_{m,M} + y_{k,K}y_{l,L}v_{m,M}\right)$$

$$= \frac{1}{2}\varepsilon_{klm}\varepsilon_{KLM}v_{k,K}y_{l,L}y_{m,M} = \frac{1}{2}v_{k,K}\varepsilon_{klm}\varepsilon_{KLM}y_{l,L}y_{m,M} \quad (1.1.49)$$

$$= \frac{1}{2}v_{k,K}2JX_{K,k} = Jv_{k,k},$$

where Eq. (1.1.22) has been used.

It can also be shown that

$$\frac{d}{dt}(dv) = v_{k,k}dv. \quad (1.1.50)$$

Proof.

$$\frac{d}{dt}(dv) = \frac{d}{dt}(JdV) = \frac{dJ}{dt}dV = Jv_{k,k}dV = v_{k,k}dv, \quad (1.1.51)$$

where Eq. (1.1.48) has been used. In addition,

$$\frac{d}{dt}(X_{L,j}) = -v_{i,K}X_{K,j}X_{L,i}. \quad (1.1.52)$$

Proof. Since

$$y_{i,K}X_{K,j} = \delta_{ij}, \quad (1.1.53)$$

we have, upon taking the material time derivative of both sides,

$$\dot{y}_{i,K}X_{K,j} + y_{i,K}\frac{d}{dt}(X_{K,j}) = 0. \quad (1.1.54)$$

Then

$$y_{i,K}\frac{d}{dt}(X_{K,j}) = -v_{i,K}X_{K,j}. \quad (1.1.55)$$

The multiplication of both sides of Eq. (1.1.55) by $X_{L,i}$ gives

$$\frac{d}{dt}(X_{L,j}) = -v_{i,K}X_{K,j}X_{L,i}. \quad (1.1.56)$$

1.2   Balance Laws

For the theory of elasticity, the relevant laws from physics are the conservation of mass, linear momentum, angular momentum and energy as follows in integral form:

$$\frac{d}{dt}\int_v \rho dv = 0, \tag{1.2.1}$$

$$\frac{d}{dt}\int_v \rho \mathbf{v} dv = \int_v \rho \mathbf{f} dv + \int_s \mathbf{t} ds, \tag{1.2.2}$$

$$\frac{d}{dt}\int_v \mathbf{y} \times \rho \mathbf{v} dv = \int_v \mathbf{y} \times \rho \mathbf{f} dv + \int_s \mathbf{y} \times \mathbf{t} ds, \tag{1.2.3}$$

$$\frac{d}{dt}\int_v \rho \left(\frac{1}{2}\mathbf{v}\cdot\mathbf{v} + \varepsilon\right) dv = \int_v \rho \mathbf{f} \cdot \mathbf{v} dv + \int_s \mathbf{t} \cdot \mathbf{v} ds, \tag{1.2.4}$$

where ρ is the mass density, $\mathbf{f}$ the body force per unit mass, $\mathbf{t}$ the surface traction per unit area, and ε the internal energy density per unit mass. We convert Eqs. (1.2.1)–(1.2.4) to differential forms as follows.

Equation (1.2.1) states that the total mass of the material body is a constant, which is the total mass in the reference state, i.e.,

$$\int_v \rho dv = \int_V \rho^0 dV, \tag{1.2.5}$$

where ρ^0 is the mass density in the reference state. For the left-hand side of Eq. (1.2.5), we apply the change of integration variables to the reference state using Eq. (1.1.41). Then the conservation of mass in Eq. (1.2.5) takes the following form:

$$\int_V (\rho J - \rho^0) dV = 0. \tag{1.2.6}$$

Equation (1.2.6) holds for any V. Assuming a continuous integrand, we conclude from Eq. (1.2.6) that

$$\rho^0 = \rho J. \tag{1.2.7}$$

Alternatively, since the total mass is conserved, we have

$$\frac{d}{dt}\int_v \rho dv = \frac{d}{dt}\int_V \rho J dV = \int_V \frac{d}{dt}(\rho J) dV$$

$$= \int_V (\dot{\rho} J + \rho \dot{J}) dV = \int_V (\dot{\rho} J + \rho J v_{i,i}) dV \tag{1.2.8}$$

$$= \int_v (\dot{\rho} + \rho v_{i,i}) dv = 0,$$

where Eq. (1.1.48) has been used. Then

$$\dot{\rho} + \rho v_{i,i} = 0. \tag{1.2.9}$$

From Eqs. (1.1.50) and (1.2.9), we can also obtain

$$\frac{d}{dt}(\rho dv) = \dot{\rho}dv + \rho\frac{d}{dt}(dv) = -\rho v_{i,i}dv + \rho v_{i,i}dv = 0. \tag{1.2.10}$$

Hence, for a differential material element of the body, the conservation of mass can be written as

$$\rho dv = \rho^0 dV. \tag{1.2.11}$$

With Eqs. (1.1.48) and (1.2.9), it can be shown that

$$\frac{d}{dt}\int_v \rho[\ ]dv = \int_v \rho\frac{d[\ ]}{dt}dv, \tag{1.2.12}$$

where the square brackets represent a tensor field in general.

Proof. With the change of integration variables to the reference state, we have

$$\begin{aligned}
\frac{d}{dt}\int_v \rho[\ ]dv &= \frac{d}{dt}\int_V \rho[\ ]JdV = \int_V \frac{d}{dt}(\rho[\ ]J)dV \\
&= \int_V \left(\dot{\rho}[\ ]J + \rho\frac{d[\ ]}{dt}J + \rho[\ ]\dot{J} \right)dV \\
&= \int_V \left(-\rho v_{i,i}[\ ]J + \rho\frac{d[\ ]}{dt}J + \rho[\ ]Jv_{i,i} \right)dV \\
&= \int_V \rho\frac{d[\ ]}{dt}JdV = \int_v \rho\frac{d[\ ]}{dt}dv.
\end{aligned} \tag{1.2.13}$$

For the conservation of linear momentum in Eq. (1.2.2), we introduce the Cauchy stress tensor $\boldsymbol{\tau}$ by

$$t_i = n_j\tau_{ji} \tag{1.2.14}$$

through the usual tetrahedron approach [1]. Then, with the use of the divergence theorem, the balance of linear momentum becomes

$$\begin{aligned}
\int_v \rho\frac{dv_i}{dt}dv &= \int_v \rho f_i dv + \int_s t_i ds \\
&= \int_v \rho f_i dv + \int_s \tau_{ji}n_j ds = \int_v \rho f_i dv + \int_v \tau_{ji,j}dv.
\end{aligned} \tag{1.2.15}$$

Hence,

$$\tau_{ji,j} + \rho f_i = \rho \dot{v}_i. \tag{1.2.16}$$

In terms of components, the balance of angular momentum in Eq. (1.2.3) takes the following form:

$$\frac{d}{dt}\int_v \varepsilon_{ijk} y_j \rho v_k dv = \int_v \varepsilon_{ijk} y_j \rho f_k dv + \int_s \varepsilon_{ijk} y_j t_k ds. \tag{1.2.17}$$

The term on the left-hand side of Eq. (1.2.17) can be written as

$$\frac{d}{dt}\int_v \varepsilon_{ijk} y_j \rho v_k dv = \int_v \rho\varepsilon_{ijk} \frac{d}{dt}(y_j v_k) dv$$

$$= \int_v \rho\varepsilon_{ijk}(\dot{y}_j v_k + y_j \dot{v}_k) dv \tag{1.2.18}$$

$$= \int_v \rho\varepsilon_{ijk}(v_j v_k + y_j \dot{v}_k) dv = \int_v \rho\varepsilon_{ijk} y_j \dot{v}_k dv.$$

The last term on the right-hand side of Eq. (1.2.17) can be written as

$$\int_s \varepsilon_{ijk} y_j t_k ds = \int_s \varepsilon_{ijk} y_j \tau_{lk} n_l ds$$

$$= \int_v (\varepsilon_{ijk} y_j \tau_{lk})_{,l} dv = \int_v \varepsilon_{ijk}(\delta_{jl}\tau_{lk} + y_j \tau_{lk,l}) dv \tag{1.2.19}$$

$$= \int_v \varepsilon_{ijk}(\tau_{jk} + y_j \tau_{lk,l}) dv.$$

Substituting Eqs. (1.2.18) and (1.2.19) into Eq. (1.2.17), we obtain

$$\int_v \rho\varepsilon_{ijk} y_j \dot{v}_k dv = \int_v \varepsilon_{ijk} y_j \rho f_k dv + \int_v \varepsilon_{ijk}(\tau_{jk} + y_j \tau_{lk,l}) dv, \tag{1.2.20}$$

or

$$\int_v \varepsilon_{ijk} y_j(\rho \dot{v}_k - \rho f_k - \tau_{lk,l}) dv = \int_v \varepsilon_{ijk} \tau_{jk} dv. \tag{1.2.21}$$

The left-hand side of Eq. (1.2.21) vanishes because of the linear momentum equation in Eq. (1.2.16). Hence,

$$\int_v \varepsilon_{ijk} \tau_{jk} dv = 0, \tag{1.2.22}$$

which implies that

$$\varepsilon_{ijk} \tau_{jk} = 0. \tag{1.2.23}$$

Equation (1.2.23) further implies that the Cauchy stress tensor τ_{kl} is symmetric, i.e., $\tau_{kl} = \tau_{lk}$.

In terms of components, the conservation of energy in Eq. (1.2.4) takes the following form:

$$\frac{d}{dt}\int_v \rho\left(\frac{1}{2}v_iv_i + \varepsilon\right) dv = \int_v \rho f_k v_k dv + \int_s t_k v_k ds. \qquad (1.2.24)$$

The left-hand side of Eq. (1.2.24) can be written as

$$\frac{d}{dt}\int_v \rho\left(\frac{1}{2}v_iv_i + \varepsilon\right) dv = \int_v \rho\frac{d}{dt}\left(\frac{1}{2}v_iv_i + \varepsilon\right) dv = \int_v \rho(v_i\dot{v}_i + \dot{\varepsilon})dv. \qquad (1.2.25)$$

The last term on the right-hand side of Eq. (1.2.24) can be written as

$$\int_s t_k v_k da = \int_s \tau_{lk} n_l v_k ds = \int_v (\tau_{lk} v_k)_{,l} dv = \int_v (\tau_{lk,l} v_k + \tau_{lk} v_{k,l})dv. \qquad (1.2.26)$$

The substitution of Eqs. (1.2.25) and (1.2.26) into Eq. (1.2.24) gives

$$\int_v \rho(v_k\dot{v}_k + \dot{\varepsilon})dv = \int_v f_k v_k dv + \int_v (\tau_{lk,l} v_k + \tau_{lk} v_{k,l})dv, \qquad (1.2.27)$$

or

$$\int_v v_k(\rho\dot{v}_k - \rho f_k - \tau_{lk,l})dv = \int_v (\tau_{lk} v_{k,l} - \rho\dot{\varepsilon})dv. \qquad (1.2.28)$$

The left-hand side of Eq. (1.2.28) vanishes because of the linear momentum equation in Eq. (1.2.16). Hence,

$$\int_v (\tau_{lk} v_{k,l} - \rho\dot{\varepsilon})dv = 0, \qquad (1.2.29)$$

which implies that

$$\rho\dot{\varepsilon} = \tau_{ij} v_{j,i}. \qquad (1.2.30)$$

The differential balance laws are obtained by applying the integral balance laws to a material body in which the fields are continuous and differentiable as many times as needed. When the material body contains a discontinuity surface, the integral balances laws lead to jump conditions [1]. We consider the special case when the discontinuity surface is an interface between two materials and assume that **y** and **v** are continuous across the interface. In this case, the conservation of mass does not lead to any jump condition [1]. In addition, the angular momentum equation and the energy equation lead to the same jump condition as the linear momentum equation [1]. Therefore, we examine the implication of the linear momentum

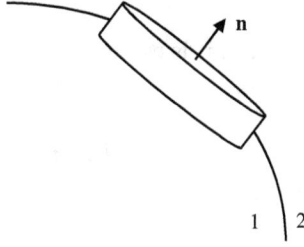

Fig. 1.2. A material interface.

equation at a material interface only. Consider the current state of an interface between medium 1 and medium 2 (see Fig. 1.2). The unit normal **n** of the interface points from medium 1 to medium 2. We construct a pillbox on the interface as shown in Fig. 1.2. The thickness of the pillbox is infinitesimal. The contributions from the volume integrals in the integral forms of the balance laws are negligible. So is the contribution from the lateral cylindrical surface. The application of the integral form of the balance of linear momentum in Eq. (1.2.2) to the pillbox yields:

$$\tau_{ij}^{(2)} n_i - \tau_{ij}^{(1)} n_i = 0, \tag{1.2.31}$$

where the top and bottom surface areas of the pillbox have canceled with each other, and higher-order infinitesimals have been neglected. Since

$$\tau_{ij} n_i ds = \tau_{ij} ds_i = \tau_{ij} J X_{L,i} dS_L = K_{Lj} dS_L = K_{Lj} N_L dS, \tag{1.2.32}$$

Eq. (1.2.31) can be written in the material form as

$$N_L K_{Lj}^{(2)} - N_L K_{Lj}^{(1)} = 0. \tag{1.2.33}$$

1.3 Constitutive Relations

For convenience, we gather the differential forms of the balance laws from Eqs. (1.2.9), (1.2.16), (1.2.23) and (1.2.30) as follows:

$$\dot{\rho} + \rho v_{i,i} = 0, \tag{1.3.1}$$

$$\rho \dot{v}_k = \tau_{ik,i} + \rho f_k, \tag{1.3.2}$$

$$\varepsilon_{ijk} \tau_{jk} = 0, \tag{1.3.3}$$

$$\rho \dot{\varepsilon} = \tau_{mk} v_{k,m}. \tag{1.3.4}$$

The above equations are written in terms of the present coordinates y_i in the sense that all spatial derivatives are taken with respect to **y**. Since in general the reference coordinates of the material points are known while the present coordinates are not, it is essential to have the equations written in terms of the reference coordinates X_K. For this purpose we introduce the first Piola–Kirchhoff stress tensor K_{Lj} which is a two-point tensor and the second Piola–Kirchhoff stress tensor P_{KL} through

$$K_{Lj} = JX_{L,i}\tau_{ij}, \quad \tau_{ij} = J^{-1}y_{i,L}K_{Lj}, \tag{1.3.5}$$

$$P_{KL} = JX_{K,i}X_{L,j}\tau_{ij}, \quad \tau_{ij} = J^{-1}y_{i,K}y_{j,L}P_{KL}. \tag{1.3.6}$$

With K_{Lj}, we have

$$\int_s t_j ds = \int_s n_i \tau_{ij} ds = \int_s \tau_{ij} ds_i$$

$$= \int_S \tau_{ij} JX_{L,i} dS_L = \int_S \tau_{ij} JX_{L,i} N_L dS \tag{1.3.7}$$

$$= \int_V (\tau_{ij} JX_{L,i})_{,L} dV = \int_V K_{Lj,L} dV.$$

Then the linear momentum equation in Eq. (1.2.2) can be written as

$$\int_V \rho \dot{v}_j J dV = \int_V \rho f_j J dV + \int_V K_{Lj,L} dV, \tag{1.3.8}$$

which implies, through the divergence theorem, that

$$K_{Lj,L} + \rho^0 f_j = \rho^0 \dot{v}_j. \tag{1.3.9}$$

The angular momentum equation does not have a spatial derivative and will not be modified. In fact, it will be satisfied automatically later when the τ_{kl} given by the constitutive relation is a symmetric tensor.

For the conservation of energy in Eq. (1.3.4), we have

$$\rho \dot{\varepsilon} = \tau_{ij} v_{j,i} = \tau_{ij}(d_{ij} + \omega_{ij}) = \tau_{ij} d_{ij}$$

$$= J^{-1} y_{i,K} y_{j,L} P_{KL} d_{ij} = J^{-1} P_{KL} \dot{E}_{KL}, \tag{1.3.10}$$

where Eq. (1.1.47) has been used. In summary, the balance laws in material form are

$$\rho^0 = \rho J, \tag{1.3.11}$$

$$K_{Lk,L} + \rho^0 f_k = \rho^0 \dot{v}_k, \tag{1.3.12}$$

$$\varepsilon_{kij}\tau_{ij} = 0, \tag{1.3.13}$$

$$\rho^0 \dot{\varepsilon} = P_{KL}\dot{E}_{KL}, \tag{1.3.14}$$

where Eq. (1.2.7) has been used and all spatial derivatives are taken with respect to $\mathbf{X}$.

The above field equations represent the relevant basic laws of physics valid for a continuum in general. The specific behavior of a particular material is specified by additional equations called constitutive relations. For constitutive relations of an elastic solid, we begin with the following internal energy density as suggested by Eq. (1.3.14):

$$\varepsilon = \varepsilon(E_{KL}). \tag{1.3.15}$$

The substitution of Eq. (1.3.15) into Eq. (1.3.14) gives

$$\left(P_{KL} - \rho^0 \frac{\partial \varepsilon}{\partial E_{KL}}\right) \dot{E}_{KL} = 0. \tag{1.3.16}$$

Since Eq. (1.3.16) is linear in $\dot{E}_{KL}$, for the equation to hold for any $\dot{E}_{KL} = \dot{E}_{LK}$, we must have

$$P_{KL} = \rho^0 \frac{1}{2} \left(\frac{\partial \varepsilon}{\partial E_{KL}} + \frac{\partial \varepsilon}{\partial E_{LK}}\right) = \rho^0 \frac{\partial \varepsilon}{\partial E_{KL}}, \tag{1.3.17}$$

where the partial derivatives with respect to E_{KL} are taken as if the strain components were independent, and ε is written as a symmetric function of E_{KL}, i.e., $\partial \varepsilon / \partial E_{KL} = \partial \varepsilon / \partial E_{LK}$. Then,

$$K_{Lj} = J X_{L,i} \tau_{ij} = J X_{L,i} (J^{-1} y_{i,K} y_{j,M} P_{KM})$$

$$= y_{j,M} P_{ML} = y_{j,K} \rho^0 \frac{\partial \varepsilon}{\partial E_{KL}}. \tag{1.3.18}$$

We also have

$$\tau_{ij} = J^{-1} y_{i,K} y_{j,L} \rho^0 \frac{\partial \varepsilon}{\partial E_{KL}} = \tau_{ji}. \tag{1.3.19}$$

The substitution Eq. (1.3.18) into Eq. (1.3.12) yields three equations for $\mathbf{y}(\mathbf{X}, t)$. Once $\mathbf{y}(\mathbf{X}, t)$ is known, the present mass density ρ can be obtained from the conservation of mass in Eq. (1.3.11).

Consider a finite body V whose boundary surface S is partitioned into S_y and S_T on which position and traction are prescribed, respectively. We assume that

$$S_y \cup S_T = S, \quad S_y \cap S_T = \emptyset. \tag{1.3.20}$$

Usual boundary-value problems consist of Eqs. (1.3.12) and (1.3.18) with the following boundary conditions:

$$y_i = \bar{y}_i \quad \text{on} \quad S_y,$$
$$N_L K_{Lk} = \bar{T}_k \quad \text{on} \quad S_T, \tag{1.3.21}$$

where $\bar{y}_i$ is the prescribed boundary position, and $\bar{T}_i$ is the surface traction per unit undeformed area. For dynamic problems, initial conditions need to be added.

1.4   Variational Formulation

Consider the following Lagrangian density L and variational functional Π:

$$L = \frac{1}{2}\rho^0 \dot{y}_i \dot{y}_i - \rho^0 \varepsilon, \tag{1.4.1}$$

$$\Pi(\mathbf{y}) = \int_{t_0}^{t_1} dt \int_V [L + \rho^0 f_i y_i] dV + \int_{t_0}^{t_1} dt \int_{S_T} \bar{T}_i y_i dS,$$

$$= \int_{t_0}^{t_1} dt \int_V \left[\frac{1}{2}\rho^0 \dot{y}_i \dot{y}_i - \rho^0 \varepsilon(\mathbf{E}) + \rho^0 f_i y_i \right] dV + \int_{t_0}^{t_1} dt \int_{S_T} \bar{T}_i y_i dS, \tag{1.4.2}$$

where

$$E_{KL} = (y_{i,K} y_{i,L} - \delta_{KL})/2. \tag{1.4.3}$$

The admissible $\mathbf{y}(\mathbf{X}, t)$ for Π satisfies the following conditions at t_0 and t_1 as well as the essential boundary conditions on S_y:

$$\delta y_i|_{t=t_0} = 0, \quad \delta y_i|_{t=t_1} = 0 \quad \text{in} \quad V,$$

$$y_i = \bar{y}_i \quad \text{on} \quad S_y, \quad t_0 < t < t_1. \tag{1.4.4}$$

Then the first variation of Π can be found as

$$\delta \Pi = \int_{t_0}^{t_1} dt \int_V (K_{Li,L} + \rho^0 f_i - \rho^0 \ddot{y}_i) \delta y_i dV$$

$$- \int_{t_0}^{t_1} dt \int_{S_T} (N_L K_{Li} - \bar{T}_i) \delta y_i dS, \tag{1.4.5}$$

where we have denoted

$$K_{Lj} = y_{j,K} \rho^0 \frac{\partial \varepsilon}{\partial E_{KL}}. \tag{1.4.6}$$

Therefore, the stationary condition of Π implies the following field equations and natural boundary conditions on S_T:

$$K_{Lk,L} + \rho^0 f_k = \rho^0 \ddot{y}_k \quad \text{in} \quad V,$$

$$N_L K_{Lk} = \bar{T}_k \quad \text{on} \quad S_T. \tag{1.4.7}$$

1.5 Third-Order Theory

The internal energy density ε that determines the constitutive relations of
nonlinear elastic materials may be written as [3]

$$
\rho^0 \varepsilon(\mathbf{E}) = \frac{1}{2} \underset{2}{c}_{ABCD} E_{AB} E_{CD} + \frac{1}{6} \underset{3}{c}_{ABCDEF} E_{AB} E_{CD} E_{EF}
$$
$$
+ \frac{1}{24} \underset{4}{c}_{ABCDEFGH} E_{AB} E_{CD} E_{EF} E_{GH} + \cdots ,
\tag{1.5.1}
$$

where a constant term which is immaterial and a term linear in $\mathbf{E}$ describing
initial stress have been dropped. The higher-order terms in Eq. (1.5.1) may
be neglected for weak nonlinearity. The material constants in Eq. (1.5.1),

$$
\underset{2}{c}_{ABCD}, \quad \underset{3}{c}_{ABCDEF}, \quad \underset{4}{c}_{ABCDEFGH},
\tag{1.5.2}
$$

are referred to as the second-, third- and fourth-order fundamental elastic
constants, respectively. The second-order constants are mainly responsible for linear material behaviors. The third- and higher-order material
constants are responsible for nonlinear behaviors. The structure of $\varepsilon(\mathbf{E})$
depends on material symmetry.

By a third-order theory, we mean that effects of all terms up to the
third powers of the displacement gradient components or their products
are included. Such a theory can be obtained by expansions and truncations
from the nonlinear theory [3]. From the internal energy density in Eq. (1.5.1)
and the constitutive relations in Eq. (1.3.18), retaining terms up to the third
order of the small displacement gradients, we obtain [3]

$$
K_{Lj} \cong \delta_{jM} \left[\underset{2}{c}_{LMAB} u_{A,B} + \frac{1}{2} \underset{2}{c}_{LMAB} u_{K,A} u_{K,B} + \underset{2}{c}_{LKAB} u_{M,K} u_{A,B} \right.
$$
$$
+ \frac{1}{2} \underset{3}{c}_{LMABCD} u_{A,B} u_{C,D} + \frac{1}{2} \underset{2}{c}_{LRAB} u_{M,R} u_{K,A} u_{K,B}
$$
$$
+ \frac{1}{2} \underset{3}{c}_{LKABCD} u_{M,K} u_{A,B} u_{C,D} + \frac{1}{2} \underset{3}{c}_{LMABCD} u_{A,B} u_{K,C} u_{K,D}
$$
$$
\left. + \frac{1}{6} \underset{4}{c}_{LMABCDEF} u_{A,B} u_{C,D} u_{E,F} \right].
\tag{1.5.3}
$$

Note that the fourth-order material constants are needed for a complete
description of the third-order effects.

As a special case of the third-order theory, if we keep terms up to the second order of the small displacement gradients only, we obtain the second-order theory as

$$K_{Lj} \cong \delta_{jM} \left[\underset{2LMAB}{c} u_{A,B} + \frac{1}{2} \underset{2LMAB}{c} u_{K,A} u_{K,B} \right.$$

$$\left. + \underset{2LKAB}{c} u_{M,K} u_{A,B} + \frac{1}{2} \underset{3LMABCD}{c} u_{A,B} u_{C,D} \right]. \qquad (1.5.4)$$

The first-order or the linear theory is given by

$$K_{Lj} \cong \delta_{jM} \underset{2LMAB}{c} u_{A,B}. \qquad (1.5.5)$$

1.6   Linear Theory for Small Deformation

To reduce the nonlinear equations of elasticity to the linear theory for infinitesimal deformations, consider small-amplitude motions of an elastic body around its reference state under small mechanical loads. It is assumed that under some norm the displacement gradient is infinitesimal, e.g.,

$$\|u_{i,K}\| \ll 1, \quad \|u_{i,K}\| = \max |u_{i,K}|. \qquad (1.6.1)$$

We want to obtain equations linear in $u_{i,M}$ and neglect their products or powers. For example,

$$\frac{\partial u_i}{\partial X_K} = \frac{\partial u_i}{\partial y_k} y_{k,K} = \frac{\partial u_i}{\partial y_k}(\delta_{kK} + u_{k,K}) \cong \frac{\partial u_i}{\partial y_k}\delta_{kK}, \qquad (1.6.2)$$

which shows that, to the lowest order of approximation, the small displacement gradients calculated from the material and spatial coordinates are approximately equal. The material time derivative of an infinitesimal field $f(\mathbf{y}, t) = f[\mathbf{y}(\mathbf{X}, t), t]$ is approximately equal to:

$$\frac{df}{dt} = \frac{\partial f}{\partial t}\bigg|_{\mathbf{X} \text{ fixed}} = \frac{\partial f}{\partial t}\bigg|_{\mathbf{y} \text{ fixed}} + \frac{\partial f}{\partial y_i}\bigg|_{t \text{ fixed}} \frac{\partial y_i}{\partial t}\bigg|_{\mathbf{X} \text{ fixed}}$$

$$= \frac{\partial f}{\partial t}\bigg|_{\mathbf{y} \text{ fixed}} + \frac{\partial f}{\partial y_i}\bigg|_{t \text{ fixed}} v_i \cong \frac{\partial f}{\partial t}\bigg|_{\mathbf{y} \text{ fixed}}. \qquad (1.6.3)$$

For the finite strain tensor, we have, approximately,

$$E_{KL} = \frac{1}{2}(u_{L,K} + u_{K,L} + u_{M,K} u_{M,L}) \cong \frac{1}{2}(u_{L,K} + u_{K,L}). \qquad (1.6.4)$$

In linear elasticity, the capital $\mathbf{X}$ is usually replaced by the lowercase $\mathbf{x}$, and only lowercase indices are used. The infinitesimal strain tensor is denoted by [4]

$$S_{kl} = \frac{1}{2}(u_{l,k} + u_{k,l}) = S_{lk}. \tag{1.6.5}$$

Similarly, we have, for the small stresses,

$$K_{Lj} \cong \delta_{Li}\tau_{ij}, \quad P_{KL} \cong \delta_{Ki}\delta_{Lj}\tau_{ij}. \tag{1.6.6}$$

Since the stress tensors are approximately the same in the linear theory, we use T_{ij} to denote the infinitesimal stress tensor [4], i.e.,

$$\tau_{ij} \cong \to T_{ij},$$
$$K_{Lj} \cong \delta_{Li}\tau_{ij} \to T_{lj}, \quad P_{KL} \cong \delta_{Ki}\delta_{Lj}\tau_{ij} \to T_{kl}. \tag{1.6.7}$$

We also introduce the internal energy density U per unit volume as follows:

$$U(\mathbf{E}) = \rho^0 \varepsilon(\mathbf{E}) \cong \frac{1}{2}\underset{2}{c}_{ABCD} E_{AB}E_{CD} \to \frac{1}{2}c_{ijkl}S_{ij}S_{kl}. \tag{1.6.8}$$

The linear constitutive relation generated by U is

$$T_{ij} = \frac{\partial U}{\partial S_{ij}} = c_{ijkl}S_{kl}. \tag{1.6.9}$$

The elastic stiffness c_{ijkl} has the following symmetries:

$$c_{ijkl} = c_{jikl} = c_{klij}. \tag{1.6.10}$$

We also assume that the elastic stiffness is positive definite in the following sense:

$$c_{ijkl}S_{ij}S_{kl} \geq 0 \quad \text{for any} \quad S_{ij} = S_{ji},$$
$$\text{and} \quad c_{ijkl}S_{ij}S_{kl} = 0 \quad \Rightarrow \quad S_{ij} = 0. \tag{1.6.11}$$

The linear constitutive relations in Eq. (1.6.9) can also be written as

$$S_{ij} = s_{ijkl}T_{kl}, \tag{1.6.12}$$

where s_{ijkl} is the elastic compliance.

In the notation of the linear theory of elasticity, the equations of motion is usually written as

$$T_{ji,j} + f_i = \rho \ddot{u}_i, \tag{1.6.13}$$

where $\mathbf{f}$ is the body force per unit volume and ρ the known mass density in the reference state which was written as ρ^0 in the nonlinear theory in the previous sections. The surface loads are also infinitesimal. We have

$$\bar{T}_j \to \bar{t}_j, \quad N_L K_{Lj} \to n_l T_{lj}. \tag{1.6.14}$$

We now introduce a compact matrix notation [4]. It allows us to use matrices to represent the stress, strain, elastic stiffness and compliance tensors and is convenient for constitutive relations of anisotropic materials. This notation consists of replacing pairs of tensor indices ij or kl by single matrix indices p or q, where i, j, k and l take the values of 1, 2 and 3, and p and q take the values of 1, 2, 3, 4, 5 and 6 according to

$$
\begin{array}{lcccccc}
ij \text{ or } kl: & 11 & 22 & 33 & 23 \text{ or } 32 & 31 \text{ or } 13 & 12 \text{ or } 21 \\
p \text{ or } q: & 1 & 2 & 3 & 4 & 5 & 6
\end{array}
\qquad (1.6.15)
$$

Thus, for the stress tensor, we write

$$
\begin{aligned}
T_1 &= T_{11}, & T_2 &= T_{22}, & T_3 &= T_{33}, \\
T_4 &= T_{23}, & T_5 &= T_{31}, & T_6 &= T_{12}.
\end{aligned}
\qquad (1.6.16)
$$

For the strain tensor, we introduce S_p such that

$$
\begin{aligned}
S_1 &= S_{11}, & S_2 &= S_{22}, & S_3 &= S_{33}, \\
S_4 &= 2S_{23}, & S_5 &= 2S_{31}, & S_6 &= 2S_{12}.
\end{aligned}
\qquad (1.6.17)
$$

Accordingly, the strain energy density per unit volume can be written as

$$
U(\mathbf{S}) = \frac{1}{2} T_p S_p = \frac{1}{2} c_{pq} S_p S_q.
\qquad (1.6.18)
$$

The corresponding constitutive relations take the following form:

$$
T_p = \frac{\partial U}{\partial S_p} = c_{pq} S_q, \quad c_{pq} = c_{qp},
$$
$$
S_p = s_{pq} T_q, \quad s_{pq} = s_{qp}.
\qquad (1.6.19)
$$

The elastic properties of a fully anisotropic material such as a triclinic crystal are represented by the following full array of $[c_{pq}]$ with 21 independent material constants:

$$
\begin{bmatrix} T_1 \\ T_2 \\ T_3 \\ T_4 \\ T_5 \\ T_6 \end{bmatrix}
=
\begin{bmatrix}
c_{11} & c_{12} & c_{13} & c_{14} & c_{15} & c_{16} \\
c_{21} & c_{22} & c_{23} & c_{24} & c_{25} & c_{26} \\
c_{31} & c_{32} & c_{33} & c_{34} & c_{35} & c_{36} \\
c_{41} & c_{42} & c_{43} & c_{44} & c_{45} & c_{46} \\
c_{51} & c_{52} & c_{53} & c_{54} & c_{55} & c_{56} \\
c_{61} & c_{62} & c_{63} & c_{64} & c_{65} & c_{66}
\end{bmatrix}
\begin{bmatrix} S_1 \\ S_2 \\ S_3 \\ S_4 \\ S_5 \\ S_6 \end{bmatrix}.
\qquad (1.6.20)
$$

Table 1.1. Elastic constants of isotropic materials.

	c_{11}, c_{12}	λ, μ	E, ν
c_{11}	c_{11}	$\lambda + 2\mu$	$\dfrac{E(1-\nu)}{(1+\nu)(1-2\nu)}$
c_{12}	c_{12}	λ	$\dfrac{E\nu}{(1+\nu)(1-2\nu)}$
λ	c_{12}	λ	$\dfrac{E\nu}{(1+\nu)(1-2\nu)}$
μ	$\dfrac{c_{11} - c_{12}}{2}$	μ	$\dfrac{E}{2(1+\nu)}$
E	$\dfrac{(c_{11} + 2c_{12})(c_{11} - c_{12})}{c_{11} + c_{12}}$	$\dfrac{\mu(3\lambda + 2\mu)}{\lambda + \mu}$	E
ν	$\dfrac{c_{12}}{c_{11} + c_{12}}$	$\dfrac{\lambda}{2(\lambda + \mu)}$	ν

In the special case of an isotropic material with two independent material constants, the stress-strain relation reduces to

$$T_1 = c_{11} S_1 + c_{12} S_2 + c_{12} S_3,$$
$$T_2 = c_{21} S_1 + c_{11} S_2 + c_{12} S_3,$$
$$T_3 = c_{21} S_1 + c_{21} S_2 + c_{11} S_3, \tag{1.6.21}$$
$$T_4 = c_{44} S_4, \quad T_5 = c_{44} S_5, \quad T_6 = c_{44} S_6,$$

where

$$c_{44} = \frac{1}{2}(c_{11} - c_{12}). \tag{1.6.22}$$

The relations of the constants c_{11} and c_{12} to Lamé's constants (λ, μ) and to Young's modulus, E, and Poisson's ratio, ν, are given in Table 1.1 [4]:

1.7 Plane Waves

For some simple problems of linear elasticity, consider plane waves in a cubic crystal of class (m3m) for which

$$[c_{pq}] = \begin{bmatrix} c_{11} & c_{12} & c_{12} & 0 & 0 & 0 \\ c_{12} & c_{11} & c_{12} & 0 & 0 & 0 \\ c_{12} & c_{12} & c_{11} & 0 & 0 & 0 \\ 0 & 0 & 0 & c_{44} & 0 & 0 \\ 0 & 0 & 0 & 0 & c_{44} & 0 \\ 0 & 0 & 0 & 0 & 0 & c_{44} \end{bmatrix}. \tag{1.7.1}$$

The stress-strain relations are

$$T_1 = c_{11}S_1 + c_{12}S_2 + c_{12}S_3,$$
$$T_2 = c_{12}S_1 + c_{11}S_2 + c_{12}S_3,$$
$$T_3 = c_{12}S_1 + c_{12}S_2 + c_{11}S_3, \qquad (1.7.2)$$
$$T_4 = c_{44}S_4, \quad T_5 = c_{44}S_5, \quad T_6 = c_{44}S_6.$$

Consider waves propagating in the x_3 direction without x_1 and x_2 dependence. They are governed by

$$c_{44}u_{1,33} = \rho\ddot{u}_1,$$
$$c_{44}u_{2,33} = \rho\ddot{u}_2, \qquad (1.7.3)$$
$$c_{11}u_{3,33} = \rho\ddot{u}_3.$$

In Eq. $(1.7.3)_1$, let

$$u_1 = U_1 \exp[i(\zeta x_3 - \omega t)], \qquad (1.7.4)$$

which is a transverse or shear wave. The substitution of Eq. (1.7.4) into Eq. $(1.7.3)_1$ leads to the following wave speed:

$$v = \frac{\omega}{\zeta} = \sqrt{\frac{c_{44}}{\rho}}. \qquad (1.7.5)$$

Equation $(1.7.3)_2$ also describes a shear wave with the same speed. For Eq. $(1.7.3)_3$, let

$$u_3 = U_3 \exp[i(\zeta x_3 - \omega t)], \qquad (1.7.6)$$

which is a longitudinal wave. The substitution of Eq. (1.7.6) into Eq. $(1.7.3)_3$ leads to the following wave speed

$$v = \frac{\omega}{\zeta} = \sqrt{\frac{c_{11}}{\rho}}. \qquad (1.7.7)$$

1.8   Thickness Vibrations of Plates

Consider a plate of cubic crystals as shown in Fig. 1.3. It has a uniform thickness $2h$ and is unbounded in the x_1 and x_2 directions. Thickness modes depend on the plate thickness coordinate x_3 and time only. The two surfaces are traction free. The boundary conditions are

$$T_{3j} = 0, \quad x_3 = \pm h. \qquad (1.8.1)$$

For cubic crystals, thickness vibrations separate into thickness-shear and thickness-stretch or thickness-extensional modes.

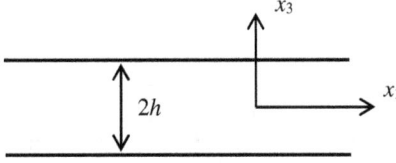

Fig. 1.3. An unbounded plate of cubic crystals.

Consider thickness-shear modes described by u_1 first. They are governed by Eqs. $(1.7.3)_1$ and $(1.8.1)$:

$$c_{44}u_{1,33} = \rho\ddot{u}_1, \quad -h < x_3 < h,$$
$$T_{31} = T_5 = c_{44}u_{1,3} = 0, \quad x_3 = \pm h. \tag{1.8.2}$$

Consider modes antisymmetric about the plate middle plane first. Let

$$u_1 = U_1 \sin\zeta x_3 \exp(i\omega t). \tag{1.8.3}$$

The boundary conditions in Eq. $(1.8.2)_2$ imply that

$$\zeta^{(n)} = \frac{n\pi}{2h}, \quad n = 1, 3, 5, \ldots. \tag{1.8.4}$$

Equation $(1.8.2)_1$ requires that

$$\omega^{(n)} = \sqrt{\frac{c_{44}}{\rho}}\zeta^{(n)} = \sqrt{\frac{c_{44}}{\rho}}\frac{n\pi}{2h}, \quad n = 1, 3, 5, \ldots. \tag{1.8.5}$$

$n = 1$ gives the fundamental thickness-shear frequency and mode. $n > 1$ are the overtone frequencies and modes. Equation $(1.8.5)$ shows that the overtones are integral multiples of the fundamental and are called harmonic overtones or harmonics. If $\cos\zeta x_3$ is used in Eq. $(1.8.3)$, a different set of modes symmetric about $x_3 = 0$ with $n = 0, 2, 4, \ldots$ will be obtained. $n = 0$ represents a rigid-body displacement and is not of interest. Static thickness-shear deformation and the first three thickness-shear modes are shown in Fig. 1.4. Thickness-shear modes described by u_2 are similar.

Thickness-stretch vibrations are governed by Eqs. $(1.7.3)_3$ and $(1.8.1)$:

$$c_{11}u_{3,33} = \rho\ddot{u}_3, \quad -h < x_3 < h,$$
$$T_{33} = T_3 = c_{11}u_{3,3} = 0, \quad x_3 = \pm h. \tag{1.8.6}$$

Let

$$u_3 = U_3 \sin\zeta x_3 \exp(i\omega t). \tag{1.8.7}$$

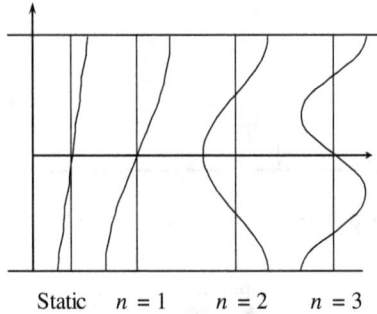

Static $n = 1$ $n = 2$ $n = 3$

Fig. 1.4. Thickness-shear deformation and modes in a plate.

The boundary conditions in Eq. (1.8.6)$_2$ imply that

$$\zeta^{(n)} = \frac{n\pi}{2h}, \quad n = 1, 3, 5, \ldots \tag{1.8.8}$$

Equation (1.8.6)$_1$ requires that

$$\omega^{(n)} = \sqrt{\frac{c_{11}}{\rho}}\zeta^{(n)} = \sqrt{\frac{c_{11}}{\rho}}\frac{n\pi}{2h}, \quad n = 1, 3, 5, \ldots \tag{1.8.9}$$

If $\cos\zeta x_3$ is used in Eq. (1.8.7), a different set of modes with $n = 0, 2, 4, \ldots$ will be obtained.

1.9 Waves in Plates

Consider the so-called antiplane or shear-horizontal (SH) waves described by $u_1 = 0, u_2 = 0$ and $u_3 = u_3(x_1, x_2, t)$ in a plate of cubic crystals as shown in Fig. 1.5. The surfaces of the plate are traction free. The governing equation is

$$T_{13,1} + T_{23,2} = c_{44}(u_{3,11} + u_{3,22}) = c_{44}\nabla^2 u_3 = \rho\ddot{u}_3, \quad -h < x_2 < h. \tag{1.9.1}$$

The boundary conditions are

$$T_{23} = c_{44}u_{3,2} = 0, \quad x_2 = \pm h. \tag{1.9.2}$$

There are two types of waves that can propagate in the plate. One is symmetric and the other antisymmetric about the plate middle plane. We discuss them separately below. For antisymmetric waves we consider the

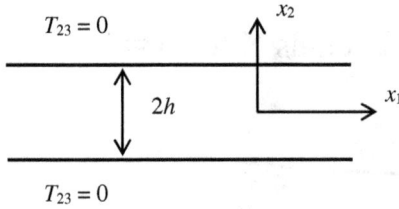

$T_{23} = 0$

$2h$

x_2

x_1

$T_{23} = 0$

Fig. 1.5. An unbounded plate of cubic crystals.

possibility of

$$u_3 = U_3 \sin \eta x_2 \cos(\xi x_1 - \omega t). \tag{1.9.3}$$

For Eq. (1.9.3) to satisfy Eq. (1.9.1), we must have

$$\eta^2 = \frac{\rho \omega^2}{c_{44}} - \xi^2 = \xi^2 \left(\frac{v^2}{v_T^2} - 1 \right),$$

$$v^2 = \frac{\omega^2}{\xi^2}, \quad v_T^2 = \frac{c_{44}}{\rho}. \tag{1.9.4}$$

The substitution of Eq. (1.9.3) into Eq. (1.9.2) leads to

$$c_{44} U_3 \eta \cos \eta h = 0. \tag{1.9.5}$$

For nontrivial solutions we must have

$$\cos \eta h = 0, \tag{1.9.6}$$

which leads to the following equation that determines the dispersion relations of the waves:

$$\eta h = \frac{n\pi}{2}, \quad n = 1, 3, 5 \ldots, \tag{1.9.7}$$

or

$$\eta^2 h^2 = \frac{\rho h^2 \omega^2}{c_{44}} - \xi^2 h^2 = \frac{n^2 \pi^2}{4}, \tag{1.9.8}$$

where Eq. (1.9.4) has been used. Equation (1.9.8) can be written into the following dimensionless form:

$$\Omega^2 - X^2 = n^2, \tag{1.9.9}$$

Ω versus X for n = 0, 1, 2, 3

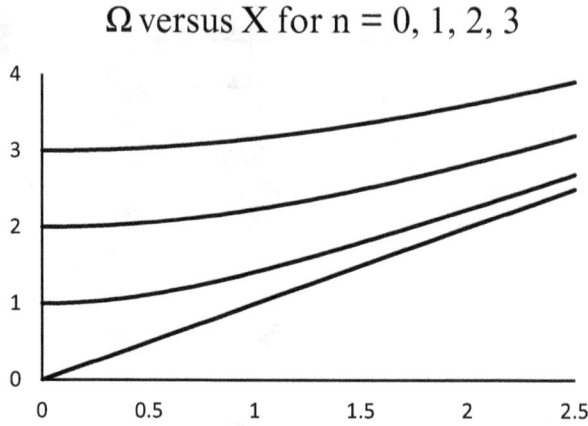

Fig. 1.6. Dispersion curves from Eq. (1.9.9) with $n = 0, 1, 2$ and 3.

where the dimensionless frequency Ω and the dimensionless wave number X in the x_1 direction are defined by

$$\Omega = \omega \left/ \left(\frac{\pi}{2h} \sqrt{\frac{c_{44}}{\rho}} \right), \quad X = \xi \left/ \left(\frac{\pi}{2h} \right). \right. \right. \tag{1.9.10}$$

For symmetric waves, we consider

$$u_3 = U_3 \cos \eta x_2 \cos(\xi x_1 - \omega t). \tag{1.9.11}$$

Through a similar analysis it also leads to Eq. (1.9.9) with n assuming even integers. The dispersion curves described by Eq. (1.9.9) are shown in Fig. 1.6.

1.10   Love Wave

Consider antiplane motions of an elastic plate on an elastic half-space of another material as shown in Fig. 1.7. The governing equations are

$$\begin{aligned} c_{44}\nabla^2 u_3 &= \rho \ddot{u}_3, \quad x_2 > 0, \\ \hat{c}_{44}\nabla^2 u_3 &= \hat{\rho} \ddot{u}_3, \quad -h < x_2 < 0, \end{aligned} \tag{1.10.1}$$

where $\hat{\rho}$ and $\hat{c}_{44}$ are the mass density and shear elastic constant of the plate. We look for solutions satisfying

$$u_3 \to 0, \quad x_2 \to +\infty. \tag{1.10.2}$$

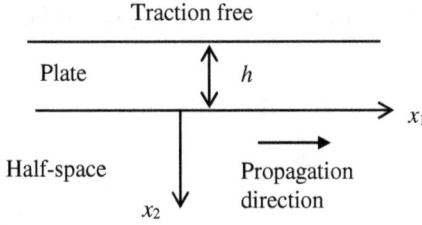

Fig. 1.7. A plate on a half-space of a different material.

For the half-space in $x_2 > 0$, we look for fields in the following form which satisfy Eq. (1.10.2):

$$u_3 = A \exp(-\eta x_2) \cos(\xi x_1 - \omega t). \qquad (1.10.3)$$

For Eq. (1.10.3) to satisfy Eq. $(1.10.1)_1$, the following must be true:

$$\eta^2 = \xi^2 - \frac{\rho \omega^2}{c_{44}} = \xi^2 \left(1 - \frac{v^2}{v_T^2} \right) > 0, \qquad (1.10.4)$$

where

$$v^2 = \frac{\omega^2}{\xi^2}, \quad v_T^2 = \frac{c_{44}}{\rho}. \qquad (1.10.5)$$

The stress component needed for the interface continuity condition is

$$T_{23} = c_{44} u_{3,2} = -c_{44} A \eta \exp(-\eta x_2) \cos(\xi x_1 - \omega t). \qquad (1.10.6)$$

For the plate in $-h < x_2 < 0$, we write

$$u_3 = (\hat{A} \cos \hat{\eta} x_2 + \hat{B} \sin \hat{\eta} x_2) \cos(\xi x_1 - \omega t). \qquad (1.10.7)$$

For Eq. (1.10.7) to satisfy Eq. $(1.10.1)_2$, we have

$$\hat{\eta}^2 = \frac{\hat{\rho} \omega^2}{\hat{c}} - \xi^2 = \xi^2 \left(\frac{v^2}{\hat{v}_T^2} - 1 \right), \qquad (1.10.8)$$

where

$$\hat{v}_T^2 = \frac{\hat{c}}{\hat{\rho}}. \qquad (1.10.9)$$

For boundary and continuity conditions, we need

$$T_{23} = \hat{c}_{44} u_{3,2} = \hat{c}_{44} (-\hat{A} \hat{\eta} \sin \hat{\eta} x_2 + \hat{B} \hat{\eta} \cos \hat{\eta} x_2) \cos(\xi x_1 - \omega t).$$
$$(1.10.10)$$

The continuity and boundary conditions are (except for a factor of $\cos(\xi x_1 - -\omega t)$):

$$u_3(0^+) = A = \hat{A} = u_3(0^-),$$

$$T_{23}(0^+) = -c_{44}A\eta = \hat{c}_{44}\hat{B}\hat{\eta} = T_{23}(0^-), \qquad (1.10.11)$$

$$T_{23}(-h) = \hat{c}_{44}(\hat{A}\hat{\eta}\sin\hat{\eta}h + \hat{B}\hat{\eta}\cos\hat{\eta}h) = 0.$$

Using Eqs. $(1.10.11)_1$ to eliminate A, we obtain

$$-c_{44}\hat{A}\eta - \hat{c}_{44}\hat{B}\hat{\eta} = 0,$$

$$\hat{A}\hat{\eta}\sin\hat{\eta}h + \hat{B}\hat{\eta}\cos\hat{\eta}h = 0. \qquad (1.10.12)$$

For nontrivial solutions,

$$\begin{vmatrix} -c_{44}\eta & -\hat{c}_{44}\hat{\eta} \\ \hat{\eta}\sin\hat{\eta}h & \hat{\eta}\cos\hat{\eta}h \end{vmatrix} = 0, \qquad (1.10.13)$$

or

$$\frac{\eta}{\xi} - \frac{\hat{c}_{44}}{c_{44}}\frac{\hat{\eta}}{\xi}\tan\hat{\eta}h = 0. \qquad (1.10.14)$$

Substituting from Eqs. (1.10.4) and (1.10.8), we obtain

$$\sqrt{1 - \frac{v^2}{v_T^2}} - \frac{\hat{c}_{44}}{c_{44}}\sqrt{\frac{v^2}{\hat{v}_T^2} - 1}\tan\left[\xi h\sqrt{\frac{v^2}{\hat{v}_T^2} - 1}\right] = 0, \qquad (1.10.15)$$

which determines the Love wave speed v as a function of the wave number ξ. The waves determined by Eq. (1.10.15) are dispersive. The dispersion relations for Love waves are real and multi-valued when $\hat{v}_T^2 < v^2 < v_T^2$, for which the shear wave speed of the plate has to be smaller than that of the half-space.

1.11   Small Deformation on a Finite Bias

Consider the following three states of an elastic body in Fig. 1.8. There are three coincident Cartesian coordinate systems. X_K are for the reference state. Greek coordinates and indices ξ_α are for the initial state. y_i are for the present state.

(i) In the reference state, the body is undeformed and free of loads. A generic point at this state is denoted by **X** with Cartesian coordinates X_K. The mass density is ρ^0.

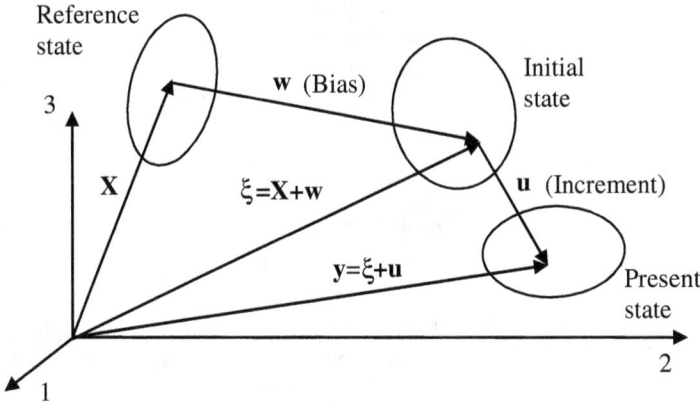

Fig. 1.8. Reference, initial and present states of an elastic body.

(ii) The initial state is static and has finite deformation. It is also called the biasing deformation. Fields of the initial state are indicated by a superscript "1". In this state the body is deformed finitely and statically under the action of body force $\mathbf{f}^1$, prescribed surface position $\bar{\boldsymbol{\xi}}$, and surface traction $\bar{T}_\alpha^1$. The displacement field from the reference state to the initial state is denoted by $\mathbf{w}(\mathbf{X})$. The position of the material point associated with $\mathbf{X}$ is given by $\boldsymbol{\xi} = \boldsymbol{\xi}(\mathbf{X})$ or $\xi_\alpha = \xi_\alpha(\mathbf{X})$. The strain at the initial state is E_{KL}^1. $\boldsymbol{\xi}(\mathbf{X})$ satisfies the following static equations of nonlinear elasticity:

$$E_{KL}^1 = (\xi_{\alpha,K}\xi_{\alpha,L} - \delta_{KL})/2, \tag{1.11.1}$$

$$P_{KL}^1 = \rho^0 \left.\frac{\partial \varepsilon}{\partial E_{KL}}\right|_{\mathbf{E}^1}, \tag{1.11.2}$$

$$K_{K\alpha}^1 = \xi_{\alpha,L} P_{KL}^1, \tag{1.11.3}$$

$$K_{K\alpha,K}^1 + \rho^0 f_\alpha^1 = 0. \tag{1.11.4}$$

(iii) In the present state, the body is under the action of time-dependent loads f_i, $\bar{y}_i$ and $\bar{T}_i$ inside the body and on its surface. The present position of the material particle associated with $\mathbf{X}$ is given by $\mathbf{y} = \boldsymbol{\xi} + \mathbf{u}$. We consider the case when the incremental displacement $\mathbf{u}$ is infinitesimal. $\mathbf{u}$ is a function of $\boldsymbol{\xi}$ and time. Since $\boldsymbol{\xi} = \boldsymbol{\xi}(\mathbf{X})$, $\mathbf{u}$ may be viewed as a function of $\mathbf{X}$ and time. Hence, $\mathbf{y}$ of the present state may also be viewed as a function of $\mathbf{X}$ and time. $\mathbf{y}(\mathbf{X}, t)$ satisfies the following dynamic equations of nonlinear elasticity:

$$E_{KL} = (y_{i,K}y_{i,L} - \delta_{KL})/2, \qquad (1.11.5)$$

$$P_{KL} = \rho^0 \frac{\partial \varepsilon}{\partial E_{KL}}\Big|_{\mathbf{E}}, \qquad (1.11.6)$$

$$K_{Lj} = y_{j,K}P_{KL}, \qquad (1.11.7)$$

$$K_{Lj,L} + \rho^0 f_j = \rho^0 \ddot{y}_j. \qquad (1.11.8)$$

Our goal is to derive linear equations governing the small and dynamic displacement $\mathbf{u}(\mathbf{X},t)$ [5,6]. To show the smallness of $\mathbf{u}$ explicitly, we write $\mathbf{y}$ as

$$y_i(\mathbf{X}, t) = \delta_{i\alpha}[\xi_\alpha(\mathbf{X}, t) + \lambda u_\alpha(\mathbf{X}, t)], \qquad (1.11.9)$$

where a dimensionless parameter λ is introduced artificially. In the following, terms quadratic in or of higher order of λ will be dropped. At the end of the derivation, λ will be set to 1. The substitution of Eq. (1.11.9) into Eq. (1.11.5) yields

$$E_{KL} \cong E^1_{KL} + \lambda \tilde{E}_{KL}, \qquad (1.11.10)$$

where the incremental strain is indicated by a superimposed tilde " $\sim$ ". In Eq. (1.11.10),

$$\tilde{E}_{KL} = (\xi_{\alpha,K}u_{\alpha,L} + \xi_{\alpha,L}u_{\alpha,K})/2 = \tilde{E}_{LK}. \qquad (1.11.11)$$

Further substitution of Eq. (1.11.10) into Eq. (1.11.6) gives

$$P_{KL} \cong P^1_{KL} + \lambda \tilde{P}_{KL}, \qquad (1.11.12)$$

where

$$\tilde{P}_{KL} = \rho^0 \frac{\partial^2 \varepsilon}{\partial E_{KL}\partial E_{MN}}\Big|_{\mathbf{E}^1} \tilde{E}_{MN}. \qquad (1.11.13)$$

Then, from Eq. (1.11.7), we have

$$K_{Ki} \cong \delta_{i\alpha}(K^1_{K\alpha} + \lambda \tilde{K}_{K\alpha}), \qquad (1.11.14)$$

where

$$\tilde{K}_{K\alpha} = u_{\alpha,L}P^1_{KL} + \xi_{\alpha,L}\tilde{P}_{KL}. \qquad (1.11.15)$$

With Eq. (1.11.13), we can write Eq. (1.11.15) as

$$\tilde{K}_{L\gamma} = G_{L\gamma M\alpha}u_{\alpha,M}, \qquad (1.11.16)$$

where

$$G_{K\alpha L\gamma} = \xi_{\alpha,M}\, \rho^0 \left.\frac{\partial^2 \varepsilon}{\partial E_{KM}\partial E_{LN}}\right|_{\mathbf{E}^1} \xi_{\gamma,N} + P^1_{KL}\delta_{\alpha\gamma} = G_{L\gamma K\alpha}. \quad (1.11.17)$$

Equation (1.11.16) is the incremental stress-displacement gradient relation. It shows that the incremental stress tensor depends linearly on the incremental displacement gradient. $G_{K\alpha L\gamma}$ are called the effective or apparent elastic constants of the material under a bias. They depend on the initial deformation $\xi_\alpha(\mathbf{X})$. Since the deformations in the present state satisfy Eq. (1.11.8) and the biasing deformations satisfy Eq. (1.11.4), we have

$$\tilde{K}_{K\alpha,K} + \rho^0 \tilde{f}_\alpha = \rho^0 \ddot{u}_\alpha, \quad (1.11.18)$$

where $\tilde{f}_\alpha$ is determined from

$$f_i = \delta_{i\alpha}(f^1_\alpha + \lambda \tilde{f}_\alpha). \quad (1.11.19)$$

In some applications, the biasing deformations are also infinitesimal and are governed by the linear theory of elasticity:

$$P^1_{KL,K} + \rho^0 f^1_L = 0, \quad (1.11.20)$$

$$P^1_{KL} = c_{KLMN}E^1_{MN}, \quad (1.11.21)$$

$$E^1_{KL} = \frac{1}{2}(w_{K,L} + w_{L,K}) = E^1_{LK}. \quad (1.11.22)$$

Since the biasing deformations are infinitesimal, we consider their first-order effects only. In this case the following internal energy density is sufficient:

$$\rho^0\varepsilon = \frac{1}{2}c_{ABCD}E_{AB}E_{CD} + \frac{1}{6}c_{ABCDEF}E_{AB}E_{CD}E_{EF}, \quad (1.11.23)$$

where we have simplified the notation of the third-order material constants and denoted

$$c_{ABCDEF} = c_{3ABCDEF}. \quad (1.11.24)$$

Then, neglecting the second-order terms of the gradients of the small $\mathbf{w}$, we can write the effective elastic constants as

$$G_{K\alpha L\gamma} = c_{K\alpha L\gamma} + \hat{c}_{K\alpha L\gamma}, \quad (1.11.25)$$

where

$$\hat{c}_{K\alpha L\gamma} = c_{K\alpha LN}w_{\gamma,N} + c_{KNL\gamma}w_{\alpha,N} + c_{K\alpha L\gamma AB}E^1_{AB} + P^1_{KL}\delta_{\alpha\gamma}. \quad (1.11.26)$$

Equation (1.11.26) shows explicitly that $G_{K\alpha L\gamma}$ depend on the small and initial deformations linearly. When the initial deformation is nonuniform, the equations for the incremental deformations are with variable coefficients. The effective elastic constants $G_{K\alpha L\gamma}$ in general have lower symmetry than the fundamental linear elastic constants $c_{K\alpha L\gamma}$. This is called induced anisotropy or symmetry breaking due to the biasing deformation. The third-order elastic constants in Eq. (1.11.26) are necessary for a complete description of the first-order effects of the biasing deformation.

1.12   Thermal and Dissipative Effects

Thermal and dissipative effects often appear together and are treated in this section [7]. For this purpose the energy equation in Eq. (1.2.4) in integral form needs to be extended to include thermal effects:

$$\frac{d}{dt} \int_v \rho \left(\frac{1}{2} \mathbf{v} \cdot \mathbf{v} + \varepsilon \right) dv = \int_v (\rho \mathbf{f} \cdot \mathbf{v} + \rho r) dv + \int_s (\mathbf{t} \cdot \mathbf{v} - \mathbf{n} \cdot \mathbf{q}) ds,$$

$$(1.12.1)$$

where $\mathbf{q}$ is the heat flux vector and r is the body heat source per unit mass. The second law of thermodynamics needs to be included as follows:

$$\frac{d}{dt} \int_v \rho \eta \, dv \geq \int_v \frac{\rho r}{\theta} dv - \int_s \frac{\mathbf{q} \cdot \mathbf{n}}{\theta} ds, \qquad (1.12.2)$$

where η is the entropy per unit mass and θ is the absolute temperature. Equations (1.12.1) and (1.12.2) can be converted to differential forms using the divergence theorem. We have

$$\rho \dot{\varepsilon} = \tau_{ij} v_{j,i} + \rho r - q_{i,i}, \qquad (1.12.3)$$

$$\rho \dot{\eta} \geq \frac{\rho r}{\theta} - \left(\frac{q_i}{\theta} \right)_{,i}. \qquad (1.12.4)$$

Eliminating r from Eqs. (1.12.3) and (1.12.4), we obtain the Clausius–Duhem inequality as

$$\rho(\theta \dot{\eta} - \dot{\varepsilon}) + \tau_{ij} v_{j,i} - \frac{q_i \theta_{,i}}{\theta} \geq 0. \qquad (1.12.5)$$

A free energy F can be introduced through the following Legendre transform:

$$F = \varepsilon - \theta \eta. \qquad (1.12.6)$$

Then the energy equation in Eq. (1.12.3) and the Clausius–Duhem inequality in Eq. (1.12.5) become

$$\rho(\dot{F} + \eta \dot{\theta} + \dot{\eta} \theta) = \tau_{ij} v_{j,i} + \rho r - q_{i,i}, \qquad (1.12.7)$$

$$-\rho(\dot{F} + \eta \dot{\theta}) + \tau_{ij} v_{j,i} - \frac{q_i \theta_{,i}}{\theta} \geq 0. \qquad (1.12.8)$$

We introduce the following material heat flux and material temperature gradient:

$$Q_K = JX_{K,k}q_k, \quad \Theta_K = \theta_{,K} = \theta_{,k}y_{k,K}. \tag{1.12.9}$$

Then Eqs. (1.12.7) and (1.12.8) can be written in material forms as

$$\rho^0(\dot{F} + \eta\dot{\theta} + \dot{\eta}\theta) = P_{KL}\dot{E}_{KL} + \rho^0 r - Q_{K,K}, \tag{1.12.10}$$

$$-\rho^0(\dot{F} + \eta\dot{\theta}) + P_{KL}\dot{E}_{KL} - \frac{Q_K\Theta_K}{\theta} \geq 0. \tag{1.12.11}$$

For constitutive relations we start with the following expressions:

$$F = F(E_{KL}; \theta; \Theta_K),$$

$$P_{KL} = P_{KL}(E_{KL}; \theta; \Theta_K; \dot{E}_{KL}), \tag{1.12.12}$$

$$Q_K = Q_K(E_{KL}; \theta; \Theta_K; \dot{E}_{KL}).$$

The substitution of Eq. (1.12.12) into the Clausius–Duhem inequality in Eq. (1.12.11) yields

$$-\rho^0\frac{\partial F}{\partial \Theta_K}\dot{\Theta}_K - \rho^0\left(\eta + \frac{\partial F}{\partial \theta}\right)\dot{\theta} + \left(P_{KL} - \rho^0\frac{\partial F}{\partial E_{KL}}\right)\dot{E}_{KL} - \frac{1}{\theta}Q_K\Theta_K \geq 0. \tag{1.12.13}$$

Since Eq. (1.12.13) is linear in $\dot{\Theta}_k$ and $\dot{\theta}$, for the inequality to hold F cannot depend on Θ_K, and η has to be related to F by

$$\eta = -\frac{\partial F}{\partial \theta}. \tag{1.12.14}$$

In addition, we break P_{KL} into recoverable and dissipative parts as follows:

$$P_{KL} = P_{KL}^R + P_{KL}^D, \tag{1.12.15}$$

where

$$P_{KL}^R = \rho^0\frac{\partial F}{\partial E_{KL}}, \quad F = F(E_{KL}; \theta), \tag{1.12.16}$$

$$P_{KL}^D = P_{KL}^D(E_{KL}; \theta; \Theta_K; \dot{E}_{KL}). \tag{1.12.17}$$

Then what is left from the Clausius–Duhem inequality in Eq. (1.12.13) is

$$P_{KL}^D\dot{E}_{KL} - \frac{1}{\theta}Q_K\Theta_K \geq 0. \tag{1.12.18}$$

From Eqs. (1.12.10) and (1.12.14), we obtain the heat or dissipation equation as

$$\rho^0\theta\dot{\eta} = P_{KL}^D\dot{E}_{KL} + \rho^0 r - Q_{K,K}, \tag{1.12.19}$$

where Eq. (1.12.16) has been used.

In summary, the nonlinear equations for thermoviscoelasticity are

$$\rho^0 = \rho J,$$

$$K_{Lk,L} + \rho^0 f_k = \rho^0 \ddot{y}_k, \qquad (1.12.20)$$

$$\rho^0 \theta \dot{\eta} = P_{KL}^D \dot{E}_{KL} + \rho^0 r - Q_{K,K},$$

with constitutive relations given by Eqs. (1.12.14)–(1.12.17) and Eq. (1.12.12)$_3$ which are restricted by Eq. (1.12.18). The equation for the conservation of mass in Eq. (1.12.20)$_1$ can be used to determine ρ separately from the other equations. Equations (1.2.20)$_{2,3}$ can be written as four equations for $y_i(\mathbf{X},t)$ and $\theta(\mathbf{X},t)$. On the boundary surface S, the thermal boundary condition may be either prescribed temperature or heat flux

$$N_L Q_L = \bar{Q}. \qquad (1.12.21)$$

1.13 Cylindrical Coordinates

To analyze circular cylindrical structures it is convenient to use cylindrical coordinates (r, θ, z) defined by

$$x_1 = r\cos\theta, \quad x_2 = r\sin\theta, \quad x_3 = z. \qquad (1.13.1)$$

In cylindrical coordinates, we have the following strain-displacement relations:

$$S_{rr} = u_{r,r}, \quad S_{\theta\theta} = \frac{1}{r}u_{\theta,\theta} + \frac{u_r}{r}, \quad S_{zz} = u_{z,z},$$

$$2S_{r\theta} = u_{\theta,r} + \frac{1}{r}u_{r,\theta} - \frac{u_\theta}{r}, \quad 2S_{\theta z} = \frac{1}{r}u_{z,\theta} + u_{\theta,z}, \qquad (1.13.2)$$

$$2S_{zr} = u_{r,z} + u_{z,r}.$$

The equations of motion take the following form:

$$\frac{\partial T_{rr}}{\partial r} + \frac{1}{r}\frac{\partial T_{\theta r}}{\partial \theta} + \frac{\partial T_{zr}}{\partial z} + \frac{T_{rr} - T_{\theta\theta}}{r} + f_r = \rho \ddot{u}_r,$$

$$\frac{\partial T_{r\theta}}{\partial r} + \frac{1}{r}\frac{\partial T_{\theta\theta}}{\partial \theta} + \frac{\partial T_{z\theta}}{\partial z} + \frac{2}{r}T_{r\theta} + f_\theta = \rho \ddot{u}_\theta, \qquad (1.13.3)$$

$$\frac{\partial T_{rz}}{\partial r} + \frac{1}{r}\frac{\partial T_{\theta z}}{\partial \theta} + \frac{\partial T_{zz}}{\partial z} + \frac{1}{r}T_{rz} + f_z = \rho \ddot{u}_z.$$

The gradient of a scalar field ψ is given by

$$\nabla \psi = \frac{\partial \psi}{\partial r}\mathbf{e}_r + \frac{1}{r}\frac{\partial \psi}{\partial \theta}\mathbf{e}_\theta + \frac{\partial \psi}{\partial z}\mathbf{e}_z. \qquad (1.13.4)$$

The divergence of a vector field $\mathbf{B}$ is

$$\nabla \cdot \mathbf{B} = \frac{1}{r}(rB_r)_{,r} + \frac{1}{r}B_{\theta,\theta} + B_{z,z}. \qquad (1.13.5)$$

The Laplace operator (Laplacian) on a scalar field ψ takes the following form:

$$\nabla^2 \psi = \frac{1}{r}\frac{\partial}{\partial r}\left(r\frac{\partial \psi}{\partial r}\right) + \frac{1}{r^2}\frac{\partial^2 \psi}{\partial \theta^2} + \frac{\partial^2 \psi}{\partial z^2}. \qquad (1.13.6)$$

Chapter 2

ELECTROMAGNETISM

This chapter presents the basics of classical electricity and magnetism [8–13]. They are about electric and magnetic fields in a vacuum or stationary and rigid materials without deformation. The spatial coordinates are x_k or $\mathbf{x}$. V, S and C represent volumes, surfaces and curves fixed in space. The international system (SI) of units is used in this chapter.

2.1 Electrostatics in a Vacuum

According to Coulomb's law between two point charges, Q and Q', the force $\mathbf{F}$ and electric field $\mathbf{E}$ on Q' at a position $\mathbf{r}$ from Q (see Fig. 2.1) are given by

$$\mathbf{F} = \frac{QQ'}{4\pi\varepsilon_0 r^3}\mathbf{r} = Q'\mathbf{E},$$

$$\mathbf{E} = \frac{Q}{4\pi\varepsilon_0}\frac{\mathbf{r}}{r^3} = \frac{Q}{4\pi\varepsilon_0}\nabla\left(\frac{-1}{r}\right),$$

$$(2.1.1)$$

where ε_0 is the electric permittivity of free-space, and we have used

$$\frac{\mathbf{r}}{r^3} = \nabla\left(\frac{-1}{r}\right). \qquad (2.1.2)$$

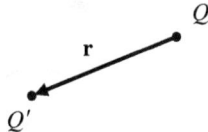

Fig. 2.1. Two point charges.

Mechanics of Ferromagnetoelastic Materials and Structures

From Eq. (2.1.1), for a closed surface S enclosing Q alone,

$$\int_S \mathbf{E} \cdot \mathbf{dS} = \int_S \frac{Q}{4\pi\varepsilon_0 r^2} \frac{\mathbf{r}}{r} \cdot \mathbf{dS}$$

$$= \int_S \frac{Q}{4\pi\varepsilon_0} d\Omega = \frac{Q}{4\pi\varepsilon_0} \int_S d\Omega = \frac{Q}{4\pi\varepsilon_0} 4\pi = \frac{Q}{\varepsilon_0}, \qquad (2.1.3)$$

where

$$\frac{1}{r^2} \frac{\mathbf{r}}{r} \cdot \mathbf{dS} = d\Omega \qquad (2.1.4)$$

has been used. $d\Omega$ is the solid angle corresponding to $\mathbf{dS}$. For a closed surface, the solid angle is 4π. The divergence of the electric field $\mathbf{E}$ in Eq. (2.1.1) is given by

$$\nabla \cdot \mathbf{E} = \frac{Q}{\varepsilon_0} \nabla \cdot \left(\frac{\mathbf{r}}{4\pi r^3} \right) = \frac{Q}{\varepsilon_0} \delta(\mathbf{r}), \qquad (2.1.5)$$

where

$$\delta(\mathbf{r}) = \frac{1}{4\pi} \nabla \cdot \left(\frac{\mathbf{r}}{r^3} \right) = \frac{1}{4\pi} \nabla \cdot \nabla \left(\frac{-1}{r} \right) = \frac{1}{4\pi} \nabla^2 \left(\frac{-1}{r} \right). \qquad (2.1.6)$$

δ is the Dirac delta function. Mathematically, Eq. (2.1.6) shows that $-1/(4\pi r)$ is the so-called fundamental solution of the Laplace operator ∇^2. By superposition, in the case of a continuous distribution of charges with density ρ^t per unit volume occupying a region V, we obtain

$$\mathbf{E}(\mathbf{x}) = \int_V \frac{\rho^t(\mathbf{x}')}{4\pi\varepsilon_0 r^2} \frac{\mathbf{r}}{r} dV', \qquad (2.1.7)$$

$$\nabla \cdot \mathbf{E} = \frac{\rho^t}{\varepsilon_0}, \qquad (2.1.8)$$

where $\mathbf{r} = \mathbf{x} - \mathbf{x}'$ as shown in Fig. 2.2, $\mathbf{x}' \in V$, and the operator ∇ is with respect to $\mathbf{x}$.

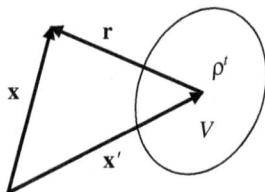

Fig. 2.2. A continuous distribution of charges in a region V.

From Eq. (2.1.1), for a closed curve C, we have

$$\oint_C \mathbf{E} \cdot \mathbf{dl} = \oint_C \frac{Q}{4\pi\varepsilon_0 r^2} \frac{\mathbf{r}}{r} \cdot \mathbf{dl}$$

$$= \oint_C \frac{Q}{4\pi\varepsilon_0} \frac{1}{r^2} dr = -\oint_C \frac{Q}{4\pi\varepsilon_0} d\left(\frac{1}{r}\right) = 0, \qquad (2.1.9)$$

where $\mathbf{dl}$ is the differential line element along the curve and

$$\frac{\mathbf{r}}{r} \cdot \mathbf{dl} = dr \qquad (2.1.10)$$

has been used. Equation (2.1.9) implies, through Stokes' theorem,

$$\nabla \times \mathbf{E} = 0. \qquad (2.1.11)$$

Then an electrostatic potential φ can be introduced such that

$$\mathbf{E} = -\nabla\varphi. \qquad (2.1.12)$$

The substitution of Eq. (2.1.12) into Eq. (2.1.8) results in a single equation for φ

$$\nabla \cdot \mathbf{E} = -\nabla \cdot (\nabla\varphi) = -\nabla^2\varphi = \frac{\rho^t}{\varepsilon_0}. \qquad (2.1.13)$$

2.2 Dielectrics

When a dielectric is placed in an electric field, the electric charges in its molecules redistribute themselves microscopically, resulting in a macroscopically polarized state. The microscopic charge redistribution occurs in different ways (see Fig. 2.3).

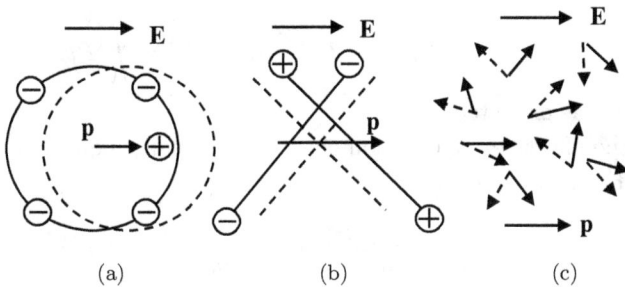

Fig. 2.3. Microscopic polarizations. (a) Electronic. (b) Ionic. (c) Orientational.

At the macroscopic level, the distinctions among different polarization mechanisms do not matter. A macroscopic polarization vector per unit volume,

$$\mathbf{P} = \lim_{\Delta V \to 0} \frac{1}{\Delta V} \sum_{\Delta V} \mathbf{p}, \qquad (2.2.1)$$

is introduced which describes the polarized state of the material macroscopically. For a dielectric body occupying a region V with a boundary surface S, the effective volume polarization charge density ρ^P and surface polarization charge density σ^P are [11]

$$\rho^P = -P_{i,i} = -\nabla \cdot \mathbf{P},$$

$$\sigma^P = n_i P_i = \mathbf{n} \cdot \mathbf{P}, \qquad (2.2.2)$$

where $\mathbf{n}$ is the outward unit normal of S. The total effective polarization charge of the body is given by

$$\int_V \rho^P dV + \int_S \sigma^P dS$$

$$= \int_V (-P_{i,i}) dV + \int_S n_i P_i dS = \int_V [(-P_{i,i}) + P_{i,i}] dV = 0, \quad (2.2.3)$$

which is as expected. The total electric moment produced by the effective polarization charges is

$$\int_V x_j \rho^P dV + \int_S x_j \sigma^P dS$$

$$= \int_V x_j (-P_{i,i}) dV + \int_S x_j n_i P_i dS$$

$$= \int_V [x_j (-P_{i,i}) + (x_j P_i)_{,i}] dV$$

$$= \int_V [x_j (-P_{i,i}) + \delta_{ij} P_i + x_j P_{i,i}] dV = \int_V P_j dV, \qquad (2.2.4)$$

which is also as expected.

With the effective polarization charges, the charge equation of electrostatics in Eq. (2.1.8) becomes

$$\nabla \cdot \mathbf{E} = E_{k,k} = \frac{\rho^t}{\varepsilon_0} = \frac{1}{\varepsilon_0} (\rho^P + \rho^e) = \frac{1}{\varepsilon_0} (-P_{k,k} + \rho^e), \qquad (2.2.5)$$

or

$$(\varepsilon_0 E_k + P_k)_{,k} = \rho^e, \qquad (2.2.6)$$

where ρ^e represents charges other than the effective polarization charges. It is usually zero in a dielectric and is kept formally only. With the introduction of an electric displacement vector $\mathbf{D}$ by

$$D_k = \varepsilon_0 E_k + P_k, \tag{2.2.7}$$

Eq. (2.2.6) can be written as

$$\nabla \cdot \mathbf{D} = D_{k,k} = \rho^e. \tag{2.2.8}$$

For a linear material,

$$P_k = \varepsilon_0 \chi^e_{kl} E_l, \tag{2.2.9}$$

where χ^e_{kl} is the electric susceptibility. Then

$$D_k = \varepsilon_0 E_k + P_k = \varepsilon_0 \delta_{kl} E_l + \varepsilon_0 \chi^e_{kl} E_l$$
$$= (\varepsilon_0 \delta_{kl} + \varepsilon_0 \chi^e_{kl}) E_l = \varepsilon_{kl} E_l, \tag{2.2.10}$$

where

$$\varepsilon_{kl} = \varepsilon_0 (\delta_{kl} + \chi^e_{kl}). \tag{2.2.11}$$

ε_{kl} is the electric permittivity tensor. In matrix notation,

$$\begin{bmatrix} D_1 \\ D_2 \\ D_3 \end{bmatrix} = \begin{bmatrix} \varepsilon_{11} & \varepsilon_{12} & \varepsilon_{13} \\ \varepsilon_{21} & \varepsilon_{22} & \varepsilon_{23} \\ \varepsilon_{31} & \varepsilon_{32} & \varepsilon_{33} \end{bmatrix} \begin{bmatrix} E_1 \\ E_2 \\ E_3 \end{bmatrix}. \tag{2.2.12}$$

The energy density of a dielectric is (see Eq. (2.9.9))

$$\hat{U}(\mathbf{D}) = \frac{1}{2} E_i D_i. \tag{2.2.13}$$

An electric enthalpy density H can be introduced through the following Legendre transform:

$$H(\mathbf{E}) = \hat{U}(\mathbf{D}) - E_i D_i = -\frac{1}{2} \varepsilon_{ij} E_i E_j. \tag{2.2.14}$$

Then Eq. (2.2.12) can be calculated from

$$D_i = -\frac{\partial H}{\partial E_i} = \varepsilon_{ik} E_k. \tag{2.2.15}$$

The electric field-potential relation is

$$\mathbf{E} = -\nabla \varphi, \quad E_i = -\varphi_{,i}. \tag{2.2.16}$$

Then

$$D_k = \varepsilon_{kl} E_l = -\varepsilon_{kl} \varphi_{,l}. \tag{2.2.17}$$

The substitution of Eq. (2.2.17) into Eq. (2.2.8) leads to a single equation for φ:

$$-(\varepsilon_{kl}\varphi_{,l})_{,k} = \rho^e. \qquad (2.2.18)$$

2.3   Conductors

In conductors there are positive charges fixed to the lattice, bound charges responsible for polarization, and free electrons that can flow through the lattice. Consider the case when a conductor is uniform and electrically neutral in a reference state. Under a voltage or electric field, the free electrons move to form currents and charge distributions. The electric field-potential relation and the charge equation of electrostatics are the same as those in Section 2.2:

$$E_k = -\varphi_{,k}, \quad D_k = \varepsilon_{kl}E_l, \qquad (2.3.1)$$

$$D_{k,k} = \rho^e. \qquad (2.3.2)$$

For a linear conductor, the current density $\mathbf{J}$ is proportional to the electric field $\mathbf{E}$ through Ohm's law

$$J_k = \sigma_{kl}E_l, \qquad (2.3.3)$$

or

$$\begin{bmatrix} J_1 \\ J_2 \\ J_3 \end{bmatrix} = \begin{bmatrix} \sigma_{11} & \sigma_{12} & \sigma_{13} \\ \sigma_{21} & \sigma_{22} & \sigma_{23} \\ \sigma_{31} & \sigma_{32} & \sigma_{33} \end{bmatrix} \begin{bmatrix} E_1 \\ E_2 \\ E_3 \end{bmatrix}, \qquad (2.3.4)$$

where σ_{kl} is the conductivity and

$$\sigma_{kl} = \sigma_{lk}. \qquad (2.3.5)$$

In terms of the electric potential φ,

$$J_k = \sigma_{kl}E_l = -\sigma_{kl}\varphi_{,l}. \qquad (2.3.6)$$

The conservation of charge or continuity equation takes the following form (see Eq. (2.9.3)):

$$\frac{\partial \rho^e}{\partial t} = -J_{k,k}, \qquad (2.3.7)$$

which, with the use of Eq. (2.3.6), becomes

$$\frac{\partial \rho^e}{\partial t} = (\sigma_{kl}\varphi_{,l})_{,k}. \qquad (2.3.8)$$

The charge equation of electrostatics in Eq. (2.3.2) can be written as

$$D_{k,k} = -(\varepsilon_{kl}\varphi_{,l})_{,k} = \rho^e. \tag{2.3.9}$$

Equations (2.3.8) and (2.3.9) are two equations for ρ^e and φ.

In the special case of an isotropic and homogeneous conductor, we have

$$\varepsilon_{kl} = \varepsilon\delta_{lk}, \quad \sigma_{kl} = \sigma\delta_{lk}, \tag{2.3.10}$$

where ε and σ are constants. Then

$$\varepsilon\nabla \cdot \mathbf{E} = \rho^e,$$
$$\dot{\rho}^e = -\nabla \cdot \mathbf{J}, \tag{2.3.11}$$
$$\mathbf{J} = \sigma\mathbf{E}.$$

With substitutions from Eqs. $(2.3.11)_{1,3}$, we can write Eq. $(2.3.11)_2$ as

$$\dot{\rho}^e = -\frac{\sigma}{\varepsilon}\rho^e. \tag{2.3.12}$$

Equation (2.3.12) can be integrated to produce

$$\rho^e(t) = \rho^e(0)\exp\left(-\frac{\sigma}{\varepsilon}t\right) = \rho^e(0)\exp\left(-\frac{t}{\tau}\right), \tag{2.3.13}$$

where

$$\tau = \frac{\varepsilon}{\sigma} \tag{2.3.14}$$

is the so-called relaxation time of a conductor which describes the time needed to reach an essentially stationary condition after an initial disturbance.

2.4 Semiconductors

In semiconductors, in addition to the effective polarization charges, there are charges from doping which are imbedded in the lattice and mobile charge carriers of holes and electrons which are responsible for the semiconduction. We assume that in the reference state the material is uniform and electrically neutral. The equations of electrostatics are

$$E_k = -\varphi_{,k},$$
$$D_k = \varepsilon_{kl}E_l, \tag{2.4.1}$$
$$D_{k,k} = q(p - n + N_D^+ - N_A^-),$$

where q is the elementary charge. p and n are the concentrations of holes and electrons. N_A^- and N_D^+ are the concentrations of ionized acceptors and donors from doping. N_A^- and N_D^+ produce holes and electrons which contribute to p and n, respectively. The continuity equations for the holes and electrons are

$$q\dot{p} = -J_{i,i}^p + \gamma^p,$$
$$q\dot{n} = J_{i,i}^n + \gamma^n, \qquad (2.4.2)$$

where $\mathbf{J}^p$ and $\mathbf{J}^n$ are the hole and electron current densities. γ^p and γ^n are the sources of holes and electrons. They may be from mechanical, thermal, electrical, magnetic and optical origins. For a macroscopic theory, specific expressions of γ^p and γ^n belong to the so-called constitutive relations. They can be determined experimentally or from microscopic theories. The constitutive relations for the current densities are

$$J_i^p = qp\mu_{ij}^p E_j - qD_{ij}^p p_{,j},$$
$$J_i^n = qn\mu_{ij}^n E_j + qD_{ij}^n n_{,j}, \qquad (2.4.3)$$

where $\boldsymbol{\mu}^p$ and $\boldsymbol{\mu}^n$ are the mobility tensors of holes and electrons, respectively. $\mathbf{D}^p$ and $\mathbf{D}^n$ are the diffusion constants. The first term in $\mathbf{J}^p$ (or $\mathbf{J}^n$) is the drift current which is nonlinear as a product of the carrier concentration p (or n) and the electric field $\mathbf{E}$. The second term in $\mathbf{J}^p$ (or $\mathbf{J}^n$) is the diffusion current. The mobility and diffusion constants satisfy the Einstein relation, e.g.,

$$\frac{\mu_{33}^p}{D_{33}^p} = \frac{\mu_{33}^n}{D_{33}^n} = \frac{q}{k_B \theta}, \qquad (2.4.4)$$

where k_B is the Boltzmann constant and θ the absolute temperature. With substitutions from Eqs. (2.4.1)$_{1,2}$ and Eq. (2.4.3), we can write Eqs. (2.4.1)$_3$ and (2.4.2) as three equations for φ, p and n.

2.5   Magnetostatics in a Vacuum

Consider a current density distribution $\mathbf{J}^t$ in a region V. For magnetostatics, $\mathbf{J}^t$ is source free with a zero divergence. According to the Biot–Savart law, the magnetic induction $\mathbf{B}$ and its force $d\mathbf{F}$ on a current element $I d\mathbf{l}$ at a point $\mathbf{x}$ is given by

$$\mathbf{B}(\mathbf{x}) = \frac{\mu_0}{4\pi} \int_V \mathbf{J}^t(\mathbf{x}') \times \frac{\mathbf{r}}{r^3} dV' = \frac{\mu_0}{4\pi} \int_V \mathbf{J}^t(\mathbf{x}') \times \nabla \left(\frac{-1}{r} \right) dV',$$
$$d\mathbf{F} = I d\mathbf{l} \times \mathbf{B}, \qquad (2.5.1)$$

where $\mathbf{x}' \in V$, $\mathbf{r} = \mathbf{x} - \mathbf{x}'$, and Eq. (2.1.2) has been used. The operator ∇ is with respect to $\mathbf{x}$. μ_0 is the magnetic permeability of free-space. Then

$$\nabla \cdot \mathbf{B} = \frac{\mu_0}{4\pi} \int_V \nabla \cdot \left\{ \mathbf{J}^t(\mathbf{x}') \times \nabla \left(\frac{-1}{r} \right) \right\} dV'$$

$$= \frac{\mu_0}{4\pi} \int_V \left\{ [\nabla \times \mathbf{J}^t(\mathbf{x}')] \cdot \frac{\mathbf{r}}{r^3} - \mathbf{J}^t(\mathbf{x}') \cdot \left[\nabla \times \nabla \left(\frac{-1}{r} \right) \right] \right\} dV'$$

$$= \frac{\mu_0}{4\pi} \int_V \{\mathbf{0} - \mathbf{0}\} dV' = 0, \qquad (2.5.2)$$

where the following vector identity [14] has been used:

$$\nabla \cdot (\mathbf{a} \times \mathbf{b}) = (\nabla \times \mathbf{a}) \cdot \mathbf{b} - \mathbf{a} \cdot (\nabla \times \mathbf{b}). \qquad (2.5.3)$$

For any scalar field f and vector field $\mathbf{a}$, we also have [14]

$$\nabla \times (f\mathbf{a}) = (\nabla f) \times \mathbf{a} + f(\nabla \times \mathbf{a}),$$
$$\nabla \cdot (f\mathbf{a}) = (\nabla f) \cdot \mathbf{a} + f(\nabla \cdot \mathbf{a}). \qquad (2.5.4)$$

According to Eq. $(2.5.4)_1$,

$$\nabla \times \left[\frac{1}{r} \mathbf{J}^t(\mathbf{x}') \right] = \left(\nabla \frac{1}{r} \right) \times \mathbf{J}^t(\mathbf{x}') + \frac{1}{r} \nabla \times \mathbf{J}^t(\mathbf{x}')$$

$$= \left(\nabla \frac{1}{r} \right) \times \mathbf{J}^t(\mathbf{x}') + \mathbf{0} = \left(\nabla \frac{1}{r} \right) \times \mathbf{J}^t(\mathbf{x}'). \qquad (2.5.5)$$

Then Eq. $(2.5.1)_1$ can be written as

$$\mathbf{B}(\mathbf{x}) = \frac{\mu_0}{4\pi} \int_V \nabla \left(\frac{1}{r} \right) \times \mathbf{J}^t(\mathbf{x}') dV' = \frac{\mu_0}{4\pi} \int_V \nabla \times \left[\frac{1}{r} \mathbf{J}^t(\mathbf{x}') \right] dV'$$

$$= \frac{\mu_0}{4\pi} \nabla \times \int_V \frac{\mathbf{J}^t(\mathbf{x}')}{r} dV' = \nabla \times \mathbf{A}, \qquad (2.5.6)$$

where we have denoted

$$\mathbf{A} = \frac{\mu_0}{4\pi} \int_V \frac{\mathbf{J}^t(\mathbf{x}')}{r} dV'. \qquad (2.5.7)$$

$\mathbf{A}$ is the vector potential of $\mathbf{B}$. Then the divergence of $\mathbf{B}$ can also be calculated from

$$\nabla \cdot \mathbf{B} = \nabla \cdot (\nabla \times \mathbf{A}) = 0. \qquad (2.5.8)$$

By another vector identity [14], the curl of $\mathbf{B}$ can be written as

$$\nabla \times \mathbf{B} = \nabla \times (\nabla \times \mathbf{A}) = \nabla(\nabla \cdot \mathbf{A}) - \nabla^2 \mathbf{A}. \qquad (2.5.9)$$

We have

$$
\begin{aligned}
\nabla \cdot \mathbf{A} &= \frac{\mu_0}{4\pi} \int_V \nabla \cdot \left[\frac{\mathbf{J}^t(\mathbf{x}')}{r} \right] dV' \\
&= \frac{\mu_0}{4\pi} \int_V \left[\nabla \left(\frac{1}{r} \right) \cdot \mathbf{J}^t(\mathbf{x}') + \frac{1}{r} \nabla \cdot \mathbf{J}^t(\mathbf{x}') \right] dV' \\
&= \frac{\mu_0}{4\pi} \int_V \left[-\nabla' \left(\frac{1}{r} \right) \cdot \mathbf{J}^t(\mathbf{x}') + 0 \right] dV' \\
&= -\frac{\mu_0}{4\pi} \int_V \nabla' \cdot \left[\mathbf{J}^t(\mathbf{x}') \frac{1}{r} \right] dV' + \frac{\mu_0}{4\pi} \int_V \frac{1}{r} \nabla' \cdot \mathbf{J}^t(\mathbf{x}') dV',
\end{aligned}
$$
$$(2.5.10)$$

where Eq. $(2.5.4)_2$ has been used twice. ∇' is with respect to $\mathbf{x}'$ and

$$\nabla' \frac{1}{r} = -\nabla \frac{1}{r}. \qquad (2.5.11)$$

The first term on the right-hand side of Eq. (2.5.10) can be converted to a surface integral using the divergence theorem. When V includes all currents, there are no currents flowing through the surface of V. Then this term vanishes. The second term on the right-hand side of Eq. (2.5.10) also vanishes because steady-state currents have a zero divergence. Therefore,

$$\nabla \cdot \mathbf{A} = 0. \qquad (2.5.12)$$

We also have

$$
\begin{aligned}
\nabla^2 \mathbf{A} &= \frac{\mu_0}{4\pi} \nabla^2 \int_V \frac{\mathbf{J}^t(\mathbf{x}')}{r} dV' = \frac{\mu_0}{4\pi} \int_V \mathbf{J}^t(\mathbf{x}') \nabla^2 \left(\frac{1}{r} \right) dV' \\
&= \frac{\mu_0}{4\pi} \int_V \mathbf{J}^t(\mathbf{x}') [-4\pi \delta(\mathbf{r})] dV' = -\mu_0 \mathbf{J}^t(\mathbf{x}),
\end{aligned}
$$
$$(2.5.13)$$

where Eq. (2.1.6) has been used. From Eqs. (2.5.9), (2.5.12) and (2.5.13), we have

$$\nabla \times \mathbf{B} = \mu_0 \mathbf{J}^t. \qquad (2.5.14)$$

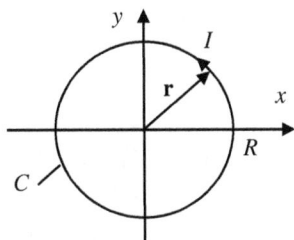

Fig. 2.4. A circular current loop.

2.6 Current Loops and Magnetic Moments

At the microscopic level, molecules carry current loops possessing magnetic moments. As a simple example, in cylindrical coordinates (r, θ, z) with unit vectors $(\mathbf{e}_r, \mathbf{e}_\theta, \mathbf{e}_z)$, the magnetic moment $\mathbf{m}$ of the circular current loop in Fig. 2.4 is defined by and found to be:

$$
\mathbf{m} = \frac{1}{2} \oint_C \mathbf{r} \times I d\mathbf{l} = \frac{1}{2} \oint_C R\mathbf{e}_r \times I dl \mathbf{e}_\theta
$$

$$
= \frac{1}{2} R I \mathbf{e}_z \oint_C dl = \frac{1}{2} R I \mathbf{e}_z 2\pi R
$$

$$
= I\pi R^2 \mathbf{e}_z = I S \mathbf{e}_z = I\mathbf{A}, \tag{2.6.1}
$$

where $A = \pi R^2$ is the area enclosed by the circle and the vector area $\mathbf{A} = A\mathbf{e}_z$. Equation (2.6.1) in the form of $\mathbf{m} = I\mathbf{A}$ is also valid for planar current loops of other shapes in general. Consider a planar current loop with area A and steady current I. In the limit when $A \to 0$, $I \to \infty$ and $IA \to m$, we have [12]

$$
\mathbf{m} = \frac{1}{2} \oint_C \mathbf{r} \times I d\mathbf{l}
$$

$$
= -\frac{1}{2} I \oint_C d\mathbf{l} \times \mathbf{r} = -\frac{1}{2} I \int_A (\mathbf{n} \times \nabla) \times \mathbf{r} dA
$$

$$
= -\frac{1}{2} I \int_A (-2\mathbf{n}) dA = I \int_A \mathbf{n} dA \to I\mathbf{A}, \tag{2.6.2}
$$

where the following vector identity has been used [14]:

$$
\oint_C d\mathbf{l} \times \mathbf{G} = \int_A (\mathbf{n} \times \nabla) \times \mathbf{G} dA. \tag{2.6.3}
$$

A current loop in a magnetic induction field $\mathbf{B}$ experiences a force $\mathbf{f}^M$ which can be calculated from the Biot–Savart law as [12]

$$d\mathbf{f}^M = I d\mathbf{l} \times \mathbf{B}, \tag{2.6.4}$$

$$\mathbf{f}^M = \oint_C I d\mathbf{l} \times \mathbf{B} = I \int_A (\mathbf{n} \times \nabla) \times \mathbf{B} dA$$

$$= I \int_A \mathbf{n} \cdot (\mathbf{B}\nabla) dA = I \left(\int_A \mathbf{n} dA \right) \cdot (\mathbf{B}\nabla) \to \mathbf{m} \cdot (\mathbf{B}\nabla), \tag{2.6.5}$$

or

$$\mathbf{f}^M = \mathbf{m} \cdot (\mathbf{B}\nabla), \quad f_j^M = m_i B_{i,j}. \tag{2.6.6}$$

The magnetic induction $\mathbf{B}$ also exerts a couple or torque on a current loop. Taking moment about a point within the area enclosed by the current loop, it can be shown that the couple acting on the current loop is given by [12]

$$\mathbf{c}^M = \oint_C \mathbf{r} \times d\mathbf{f}^M = \oint_C \mathbf{r} \times (I d\mathbf{l} \times \mathbf{B}) = \mathbf{m} \times \mathbf{B}. \tag{2.6.7}$$

In Eq. (2.6.7), the component of $\mathbf{B}$ along $\mathbf{m}$ does not contribute to $\mathbf{c}^M$.

For the fields produced by $\mathbf{m}$ in a vacuum, it can be shown that [12, 13]

$$\mathbf{A} = \frac{\mu_0}{4\pi} \mathbf{m} \times \frac{\mathbf{r}}{r^3},$$

$$\mathbf{B} = -\frac{\mu_0}{4\pi} (\mathbf{m} \cdot \nabla) \frac{\mathbf{r}}{r^3}. \tag{2.6.8}$$

2.7 Magnetization

In a material, the microscopic magnetic moments may be randomly oriented as shown in Fig. 2.5 or aligned to various degrees for different reasons. We

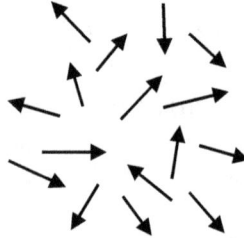

Fig. 2.5. Microscopic magnetic moments in matter.

define a macroscopic magnetization vector **M** per unit volume by

$$\mathbf{M} = \lim_{\Delta V \to 0} \frac{1}{\Delta V} \sum_{\Delta V} \mathbf{m}. \qquad (2.7.1)$$

For a finite body occupying a volume V whose surface is S with an outward unit normal **n**, it can be shown [12] that effectively **M** is equivalent to the following volume magnetization current density $\mathbf{J}^M$ and surface magnetization current density $\mathbf{j}^M$:

$$\mathbf{J}^M = \nabla \times \mathbf{M}$$

$$= \left(\frac{\partial M_z}{\partial y} - \frac{\partial M_y}{\partial z}\right) \mathbf{i} + \left(\frac{\partial M_x}{\partial z} - \frac{\partial M_z}{\partial x}\right) \mathbf{j} + \left(\frac{\partial M_y}{\partial x} - \frac{\partial M_x}{\partial y}\right) \mathbf{k}, \qquad (2.7.2)$$

$$\mathbf{j}^M = \mathbf{M} \times \mathbf{n}.$$

For example, part of the z (or **k**) component of Eq. (2.7.2) when the variation of M_x along y is considered can be calculated from Fig. 2.6 as follows. We have, from the figure,

$$I' \Delta y \Delta z = M_x \Delta x \Delta y \Delta z, \qquad (2.7.3)$$

$$I'' \Delta y \Delta z = \left(M_x + \frac{\partial M_x}{\partial y} \Delta y\right) \Delta x \Delta y \Delta z. \qquad (2.7.4)$$

Then

$$I' - I'' = -\frac{\partial M_x}{\partial y} \Delta x \Delta y = J_z^M \Delta x \Delta y, \qquad (2.7.5)$$

or

$$J_z^M = -\frac{\partial M_x}{\partial y}. \qquad (2.7.6)$$

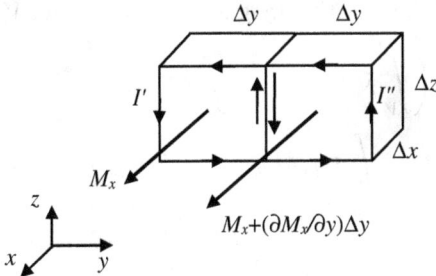

Fig. 2.6. Contributions to the z component of $\mathbf{J}^M$ from M_x.

Similarly, when the variation of M_y along x is considered, we have

$$J_z^M = \frac{\partial M_y}{\partial x}. \tag{2.7.7}$$

Adding Eqs. (2.7.6) and (2.7.7), we obtain the z component of the $\mathbf{J}^M$ in Eq. (2.7.2):

$$J_z^M = \frac{\partial M_y}{\partial x} - \frac{\partial M_x}{\partial y}. \tag{2.7.8}$$

With Eq. (2.7.2), we can write Eq. (2.5.14) as

$$\nabla \times \mathbf{B} = \mu_0(\mathbf{J} + \mathbf{J}^M), \tag{2.7.9}$$

or

$$\nabla \times \left(\frac{1}{\mu_0}\mathbf{B} - \mathbf{M}\right) = \mathbf{J}, \tag{2.7.10}$$

where $\mathbf{J} = \mathbf{J}^t - \mathbf{J}^M$ is the true current density [9]. With the introduction of the magnetic field vector $\mathbf{H}$ by

$$\mathbf{H} = \frac{1}{\mu_0}\mathbf{B} - \mathbf{M}, \tag{2.7.11}$$

or

$$\mathbf{B} = \mu_0(\mathbf{H} + \mathbf{M}), \tag{2.7.12}$$

Eq. (2.7.10) can be written as

$$\nabla \times \mathbf{H} = \mathbf{J}. \tag{2.7.13}$$

The integration of the effective magnetic currents over V and S together is

$$\left(\int_V \nabla \times \mathbf{M} dV + \int_S \mathbf{M} \times \mathbf{n}\right)_i$$
$$= \int_V \varepsilon_{ijk} M_{k,j} dV + \int_S \varepsilon_{ijk} M_j n_k dS$$
$$= \int_V \varepsilon_{ijk} M_{k,j} dV + \int_V \varepsilon_{ijk} M_{j,k} dV$$
$$= \int_V \varepsilon_{ijk} (M_{k,j} + M_{j,k}) dV = 0. \tag{2.7.14}$$

The total force on the effective magnetic currents over V and S together is

$$\int_V \varepsilon_{ijk} J_j^M B_k dV + \int_S \varepsilon_{ijk} j_j^M B_k dS$$

$$= \int_V \varepsilon_{ijk} \varepsilon_{jmn} M_{n,m} B_k dV + \int_S \varepsilon_{ijk} \varepsilon_{jmn} M_m n_n B_k dS$$

$$= \int_V \varepsilon_{jki} \varepsilon_{jmn} M_{n,m} B_k dV + \int_V \varepsilon_{jki} \varepsilon_{jmn} (M_m B_k)_{,n} dS$$

$$= \int_V M_k B_{k,i} dV = \int_V f_i^M dV, \tag{2.7.15}$$

where, similar to Eq. (2.6.6), the magnetic force on the magnetization $\mathbf{M}$ per unit volume is [12]

$$\mathbf{f}^M = \mathbf{M} \cdot (\mathbf{B}\nabla), \quad f_j^M = M_i B_{i,j}. \tag{2.7.16}$$

It can be verified that $\mathbf{f}^M + \mathbf{J} \times \mathbf{B}$ can be written as the divergence of a tensor $\mathbf{T}^M$ as [12]

$$f_l^M + (\mathbf{J} \times \mathbf{B})_l = T_{ml,m}^M, \tag{2.7.17}$$

where

$$T_{ml}^M = B_m H_l + \frac{\mu_0}{2}(M_k M_k - H_k H_k)\delta_{ml}. \tag{2.7.18}$$

2.8 Linear Magnetic Materials

For a linear material,

$$M_k = \chi_{kl}^M H_l, \tag{2.8.1}$$

where χ_{kl}^M is the magnetic susceptibility. Then

$$B_k = \mu_0(H_k + M_k) = \mu_0(\delta_{kl} H_l + \chi_{kl}^M H_l) = \mu_{kl} H_l, \tag{2.8.2}$$

or

$$\begin{bmatrix} B_1 \\ B_2 \\ B_3 \end{bmatrix} = \begin{bmatrix} \mu_{11} & \mu_{12} & \mu_{13} \\ \mu_{21} & \mu_{22} & \mu_{23} \\ \mu_{31} & \mu_{32} & \mu_{33} \end{bmatrix} \begin{bmatrix} H_1 \\ H_2 \\ H_3 \end{bmatrix}, \tag{2.8.3}$$

where

$$\mu_{kl} = \mu_0(\delta_{kl} + \chi_{kl}^M) \tag{2.8.4}$$

is the magnetic permeability tensor. For a linear magnetic material, the energy density per unit volume is (see Eq. (2.9.9))

$$\hat{U}(\mathbf{B}) = \frac{1}{2}H_i B_i. \tag{2.8.5}$$

With the following Legendre transform, we introduce a magnetic enthalpy function H per unit volume by

$$H(\mathbf{H}) = \hat{U}(\mathbf{B}) - H_i B_i = -\frac{1}{2}H_i B_i = -\frac{1}{2}\mu_{ij}H_i H_j. \tag{2.8.6}$$

Then

$$B_i = -\frac{\partial H}{\partial H_i} = \mu_{ik} H_k. \tag{2.8.7}$$

In a region where $\mathbf{J} = 0$, Eq. (2.7.13) reduces to

$$\nabla \times \mathbf{H} = 0. \tag{2.8.8}$$

Then a scalar potential ψ can be introduced such that

$$\mathbf{H} = -\nabla\psi, \quad H_i = -\psi_{,i}. \tag{2.8.9}$$

With ψ, Eq. (2.8.7) becomes

$$B_k = \mu_{kl}H_l = -\mu_{kl}\psi_{,l}, \tag{2.8.10}$$

which leads to the following equation for ψ:

$$\nabla \cdot \mathbf{B} = B_{k,k} = (\mu_{kl}H_l)_{,k} = -(\mu_{kl}\psi_{,l})_{,k} = 0. \tag{2.8.11}$$

2.9   Maxwell's Equations

For time-dependent problems, electric and magnetic fields are coupled dynamically and are governed by Maxwell's equations:

$$\nabla \cdot \mathbf{D} = \rho^e,$$

$$\nabla \cdot \mathbf{B} = 0,$$

$$\nabla \times \mathbf{E} = -\frac{\partial \mathbf{B}}{\partial t}, \tag{2.9.1}$$

$$\nabla \times \mathbf{H} = \mathbf{J} + \frac{\partial \mathbf{D}}{\partial t}.$$

Taking the time derivative of the first equation and the divergence of the fourth equation in Eq. (2.9.1), respectively, we obtain

$$\frac{\partial}{\partial t}(\nabla \cdot \mathbf{D}) = \frac{\partial \rho^e}{\partial t},$$

$$0 = \nabla \cdot \mathbf{J} + \frac{\partial}{\partial t}(\nabla \cdot \mathbf{D}). \tag{2.9.2}$$

Eliminating the time derivative of the divergence of $\mathbf{D}$, we obtain the conservation of charge as

$$\frac{\partial \rho^e}{\partial t} = -\nabla \cdot \mathbf{J}. \tag{2.9.3}$$

Taking the scalar products of the third and fourth equations of Eq. (2.9.1) with $\mathbf{H}$ and $\mathbf{E}$, respectively, we have

$$\mathbf{H} \cdot (\nabla \times \mathbf{E}) = -\mathbf{H} \cdot \frac{\partial \mathbf{B}}{\partial t},$$
$$\mathbf{E} \cdot (\nabla \times \mathbf{H}) = \mathbf{E} \cdot \mathbf{J} + \mathbf{E} \cdot \frac{\partial \mathbf{D}}{\partial t}. \tag{2.9.4}$$

Subtracting the two equations in Eq. (2.9.4) from each other and using the vector identity

$$\nabla \cdot (\mathbf{E} \times \mathbf{H}) = \mathbf{H} \cdot (\nabla \times \mathbf{E}) - \mathbf{E} \cdot (\nabla \times \mathbf{H}), \tag{2.9.5}$$

we obtain Poynting's theorem as

$$\mathbf{E} \cdot \frac{\partial \mathbf{D}}{\partial t} + \mathbf{H} \cdot \frac{\partial \mathbf{B}}{\partial t} + \mathbf{E} \cdot \mathbf{J} = -\nabla \cdot (\mathbf{E} \times \mathbf{H}), \tag{2.9.6}$$

or

$$\frac{\partial \hat{U}}{\partial t} + \mathbf{E} \cdot \mathbf{J} = -\nabla \cdot \mathbf{S}, \tag{2.9.7}$$

where

$$\frac{\partial \hat{U}}{\partial t} = \mathbf{E} \cdot \frac{\partial \mathbf{D}}{\partial t} + \mathbf{H} \cdot \frac{\partial \mathbf{B}}{\partial t}, \quad \mathbf{S} = \mathbf{E} \times \mathbf{H},$$
$$\hat{U} = \hat{U}(\mathbf{D}, \mathbf{B}), \quad \mathbf{E} = \frac{\partial \hat{U}}{\partial \mathbf{D}}, \quad \mathbf{H} = \frac{\partial \hat{U}}{\partial \mathbf{B}}. \tag{2.9.8}$$

$\hat{U}$ does not depend on t explicitly. It is a composite function of t through $\mathbf{D}$ and $\mathbf{B}$. $\mathbf{S}$ is the Poynting vector or the electromagnetic energy flux per unit area and unit time. For a linear material,

$$\hat{U} = \frac{1}{2}\mathbf{E} \cdot \mathbf{D} + \frac{1}{2}\mathbf{H} \cdot \mathbf{B}. \tag{2.9.9}$$

From Eq. (2.9.8), using

$$\mathbf{D} = \varepsilon_0 \mathbf{E} + \mathbf{P}, \quad \mathbf{H} = \frac{\mathbf{B}}{\mu_0} - \mathbf{M}, \tag{2.9.10}$$

we have

$$\frac{\partial \hat{U}}{\partial t} = \mathbf{E} \cdot \left(\varepsilon_0 \frac{\partial \mathbf{E}}{\partial t} + \frac{\partial \mathbf{P}}{\partial t} \right) + \left(\frac{\mathbf{B}}{\mu_0} - \mathbf{M} \right) \cdot \frac{\partial \mathbf{B}}{\partial t}$$

$$= \mathbf{E} \cdot \varepsilon_0 \frac{\partial \mathbf{E}}{\partial t} + \frac{\mathbf{B}}{\mu_0} \cdot \frac{\partial \mathbf{B}}{\partial t} + \mathbf{E} \cdot \frac{\partial \mathbf{P}}{\partial t} - \mathbf{M} \cdot \frac{\partial \mathbf{B}}{\partial t}$$

$$= \frac{\partial}{\partial t} \left(\frac{\varepsilon_0}{2} \mathbf{E} \cdot \mathbf{E} + \frac{1}{2\mu_0} \mathbf{B} \cdot \mathbf{B} \right) + \mathbf{E} \cdot \frac{\partial \mathbf{P}}{\partial t} - \mathbf{M} \cdot \frac{\partial \mathbf{B}}{\partial t}, \quad (2.9.11)$$

or

$$\frac{\partial \hat{U}}{\partial t} = \frac{\partial U^F}{\partial t} + \frac{\partial U}{\partial t}, \quad (2.9.12)$$

where

$$U^F = \frac{\varepsilon_0}{2} \mathbf{E} \cdot \mathbf{E} + \frac{1}{2\mu_0} \mathbf{B} \cdot \mathbf{B},$$

$$\frac{\partial U}{\partial t} = \mathbf{E} \cdot \frac{\partial \mathbf{P}}{\partial t} - \mathbf{M} \cdot \frac{\partial \mathbf{B}}{\partial t}. \quad (2.9.13)$$

$\hat{U}$ includes the field energy density U^F and the internal energy density U due to polarization and magnetization. They are all per unit volume.

As an example of the applications of Maxwell's equations, consider electromagnetic waves in a vacuum. Equation (2.9.1) reduces to

$$\nabla \times \mathbf{E} = -\frac{\partial \mathbf{B}}{\partial t},$$

$$\nabla \times \mathbf{H} = \frac{\partial \mathbf{D}}{\partial t}, \quad (2.9.14)$$

$$\nabla \cdot \mathbf{D} = 0,$$

$$\nabla \cdot \mathbf{B} = 0.$$

For a vacuum,

$$\mathbf{D} = \varepsilon_0 \mathbf{E}, \quad \mathbf{B} = \mu_0 \mathbf{H}. \quad (2.9.15)$$

Taking the curl of Eq. $(2.9.14)_1$, we have

$$\nabla \times (\nabla \times \mathbf{E}) = -\nabla \times \frac{\partial \mathbf{B}}{\partial t}. \quad (2.9.16)$$

With the use of the following vector identity

$$\nabla \times (\nabla \times \mathbf{E}) = \nabla(\nabla \cdot \mathbf{E}) - \nabla^2 \mathbf{E}, \quad (2.9.17)$$

Eq. (2.9.16) becomes

$$\nabla(\nabla \cdot \mathbf{E}) - \nabla^2 \mathbf{E} = -\frac{\partial}{\partial t}(\nabla \times \mathbf{B}), \quad (2.9.18)$$

or

$$\frac{1}{\varepsilon_0} \nabla (\nabla \cdot \mathbf{D}) - \nabla^2 \mathbf{E} = -\frac{\partial}{\partial t} (\nabla \times \mu_0 \mathbf{H}), \qquad (2.9.19)$$

where Eq. (2.9.15) has been used. Using Eqs. (2.9.14)$_{2,3}$, we obtain, from Eq. (2.9.19),

$$\mathbf{0} - \nabla^2 \mathbf{E} = -\mu_0 \frac{\partial}{\partial t} \frac{\partial \mathbf{D}}{\partial t}, \qquad (2.9.20)$$

or

$$\nabla^2 \mathbf{E} = \varepsilon_0 \mu_0 \frac{\partial^2 \mathbf{E}}{\partial t^2}. \qquad (2.9.21)$$

Equation (2.9.21) can be written as

$$\nabla^2 \mathbf{E} = \frac{1}{c^2} \frac{\partial^2 \mathbf{E}}{\partial t^2}, \qquad (2.9.22)$$

which is the standard wave equation where

$$c = \frac{1}{\sqrt{\varepsilon_0 \mu_0}} \qquad (2.9.23)$$

is the wave speed which in this case is the speed of light in a vacuum. Similarly, it can be shown that

$$\nabla^2 \mathbf{B} = \frac{1}{c^2} \frac{\partial^2 \mathbf{B}}{\partial t^2}. \qquad (2.9.24)$$

The power of electromagnetic fields on a current loop is given by [12, 15]

$$w^M = \oint_C I \mathbf{dl} \cdot \mathbf{E} = \int_A I \mathbf{n} \cdot (\nabla \times \mathbf{E}) dA$$

$$= -\int_A I \mathbf{n} \cdot \frac{\partial \mathbf{B}}{\partial t} dA = -\left(\int_A I \mathbf{n} dA \right) \cdot \frac{\partial \mathbf{B}}{\partial t} \rightarrow -\mathbf{m} \cdot \frac{\partial \mathbf{B}}{\partial t}, \qquad (2.9.25)$$

where

$$\nabla \times \mathbf{E} = -\frac{\partial \mathbf{B}}{\partial t} \qquad (2.9.26)$$

has been used.

Chapter 3

PIEZOMAGNETIC AND MAGNETOSTRICTIVE EFFECTS

This chapter is about basic magnetoelastic couplings in insulators, such as piezomagnetic and magnetostrictive effects. A reduced version of the two-continuum model of [16–19] is used to construct the nonlinear theoretical framework of this chapter. Different from [16–19] where the magnetic field **H** is used to express the magnetic force and couple on a magnetic moment, the magnetic induction **B** is used in this chapter, which is the same as the previous chapter and what is in the later paper [15] by the author of [16–19]. The reason of using **B** instead of **H** is because of the current loop model of magnetic moments in the previous chapter where the force and couple on a current loop were calculated from the Biot–Savart law in terms of **B**. The SI units are used in this chapter. Electric polarization is not considered in this chapter. It will be included in the general theory in the last chapter of this book.

3.1 Two-Continuum Model

The two continua are the lattice continuum and the effective circulating current continuum shown in Fig. 3.1 where, for simplicity, the circulating current continuum is denoted by the spin continuum in the figure. The two continua are shown in Fig. 3.1 separately along with the mechanical loads, magnetic loads and the interactions between the two continua. The combination of the two continua together is shown in Fig. 3.2 with external loads but not interactions. The spin continuum is assumed to be massless. Hence, it does not have linear and angular momenta. The spin continuum cannot move relative to the lattice continuum, but its magnetic moment vector can change its magnitude and direction. **f** and **t** are the usual mechanical body force and surface traction on the lattice. $\mathbf{f}^L$ is a local force

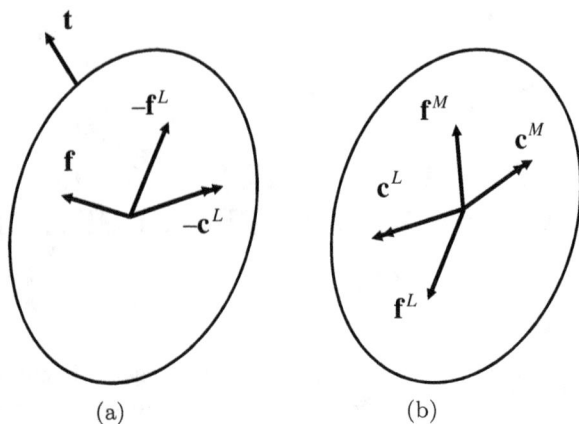

Fig. 3.1.   Separate lattice and spin continua: (a) lattice; (b) spin.

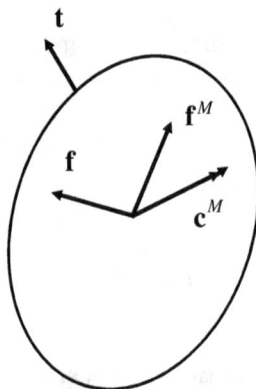

Fig. 3.2.   Combined continuum of lattice and spin.

representing the interaction between the two continua. The two continua also interact with a local couple $\mathbf{c}^L$. $\mathbf{f}^M$ is the magnetic body force on the spin continuum due to the Maxwellian magnetic induction $\mathbf{B}^M$. From Section 2.7,

$$f_i^M = M_k B_{k,i}^M, \quad \mathbf{f}^M = \mathbf{M} \cdot (\mathbf{B}^M \nabla), \tag{3.1.1}$$

$$f_l^M = T_{ml,m}^M, \tag{3.1.2}$$

$$T_{ml}^M = B_m H_l + \frac{\mu_0}{2} \left(M_k M_k - H_k H_k \right) \delta_{ml}. \tag{3.1.3}$$

$\mathbf{B}^M$ also produces a body couple $\mathbf{c}^M$ on the spin continuum. From Section 2.6, $\mathbf{c}^M$ has the following expression:

$$\mathbf{c}^M = \mathbf{M} \times \mathbf{B}^M, \quad c_i^M = \varepsilon_{ijk} M_j B_k^M. \tag{3.1.4}$$

For later convenience, we introduce the magnetization per unit mass by

$$\boldsymbol{\mu} = \frac{\mathbf{M}}{\rho}. \tag{3.1.5}$$

Corresponding to Eq. (2.9.25), for the magnetic power per unit volume, we have

$$
\begin{aligned}
w^M &= -M_j \frac{\partial B_j^M}{\partial t} = -\rho \mu_j \frac{\partial B_j^M}{\partial t} \\
&= -\rho \mu_j \frac{\partial B_j^M}{\partial t} - \rho \mu_j v_i B_{j,i}^M + \rho \mu_j v_i B_{j,i}^M \\
&= -\rho \mu_j \frac{d B_j^M}{dt} + f_i^M v_i.
\end{aligned}
\tag{3.1.6}
$$

3.2 Integral Balance Laws

Let $\mathbf{X}$ and $\mathbf{y}$ be the reference and present coordinates of a material point of the lattice continuum. l, s and v represent lines, surfaces and volumes. The integral forms of the balance laws for the quasistatic magnetic fields are

$$\int_s \mathbf{n} \cdot \mathbf{B}^M \, ds = 0, \tag{3.2.1}$$

$$\int_l \mathbf{H} \cdot \mathbf{dl} = 0. \tag{3.2.2}$$

For the lattice and spin continua together in Fig. 3.2, the conservation of mass, linear momentum, angular momentum and energy in integral forms are

$$\frac{d}{dt} \int_v \rho \, dv = 0, \tag{3.2.3}$$

$$\frac{d}{dt} \int_v \rho \mathbf{v} \, dv = \int_s \mathbf{t} \, ds + \int_v (\rho \mathbf{f} + \mathbf{f}^M) \, dv, \tag{3.2.4}$$

$$\frac{d}{dt} \int_v \mathbf{y} \times \rho \mathbf{v} \, dv = \int_s \mathbf{y} \times \mathbf{t} \, ds + \int_v [\mathbf{y} \times (\rho \mathbf{f} + \mathbf{f}^M) + \mathbf{c}^M] \, dv, \tag{3.2.5}$$

$$\frac{d}{dt} \int_v \rho \left(\frac{1}{2} \mathbf{v} \cdot \mathbf{v} + \varepsilon \right) dv = \int_s \mathbf{t} \cdot \mathbf{v} \, ds + \int_v (\rho \mathbf{f} \cdot \mathbf{v} + w^M) \, dv. \tag{3.2.6}$$

3.3   Differential Balance Laws

The differential forms of Eqs. (3.2.1)–(3.2.5) are

$$B_{k,k}^M = 0, \tag{3.3.1}$$

$$\varepsilon_{ijk}H_{k,j} = 0 \quad \text{or} \quad H_k = -\psi_{,k}, \tag{3.3.2}$$

$$\rho^0 = \rho J, \tag{3.3.3}$$

$$\tau_{ij,i} + \rho f_j + f_j^M = \rho \ddot{u}_j, \tag{3.3.4}$$

$$\varepsilon_{ijk}\tau_{jk} + c_i^M = 0. \tag{3.3.5}$$

The energy equation in Eq. (3.2.6) can be brought into differential form as

$$\rho \frac{d\varepsilon}{dt} = \tau_{ij}v_{j,i} - f_j^M v_j - M_j \frac{\partial B_j^M}{\partial t}, \tag{3.3.6}$$

where the linear momentum equation has been used. With the use of Eq. (3.1.6), we can write Eq. (3.3.6) as

$$\rho \frac{d\varepsilon}{dt} = \tau_{ij}v_{j,i} - M_j \frac{dB_j^M}{dt}. \tag{3.3.7}$$

When applied to an interface between two materials or the boundary surface of a finite body, the integral balance laws lead to jump conditions or boundary conditions [16–19]. On a boundary surface, we may prescribe

$$\mathbf{y} \quad \text{or} \quad \mathbf{n} \cdot \boldsymbol{\tau}, \tag{3.3.8}$$

$$\psi \quad \text{or} \quad \mathbf{n} \cdot \mathbf{B}^M. \tag{3.3.9}$$

3.4   Constitutive Relations

We introduce an enthalpy function χ through

$$\chi = \varepsilon + B_i^M \mu_i. \tag{3.4.1}$$

Then the energy equation in Eq. (3.3.7) becomes

$$\rho \frac{d\chi}{dt} = \tau_{ij}v_{j,i} + B_j^M \rho \frac{d\mu_j}{dt} \tag{3.4.2}$$

or

$$\rho \frac{d\chi}{dt} = \tau_{ij}X_{M,i}\frac{d}{dt}(y_{j,M}) + \rho B_j^M \frac{d\mu_j}{dt}, \tag{3.4.3}$$

where we have used

$$v_{j,i} = X_{M,i}\frac{d}{dt}(y_{j,M}). \tag{3.4.4}$$

According to Eq. (3.4.3), we let

$$\chi = \chi(y_{j,M}; \mu_i). \tag{3.4.5}$$

Then,

$$\frac{d\chi}{dt} = \frac{\partial\chi}{\partial(y_{j,M})}\frac{d}{dt}(y_{j,M}) + \frac{\partial\chi}{\partial\mu_i}\frac{d\mu_i}{dt}. \tag{3.4.6}$$

Substituting Eq. (3.4.6) into Eq. (3.4.3), we obtain

$$\rho\frac{\partial\chi}{\partial(y_{j,M})}\frac{d}{dt}(y_{j,M}) + \rho\frac{\partial\chi}{\partial\mu_i}\frac{d\mu_i}{dt} = \tau_{ij}X_{M,i}\frac{d}{dt}(y_{j,M}) + \rho B_j^M\frac{d\mu_j}{dt} \tag{3.4.7}$$

or

$$\left[X_{M,i}\tau_{ij} - \rho\frac{\partial\chi}{\partial(y_{j,M})}\right]\frac{d}{dt}(y_{j,M}) + \rho\left[B_i^M - \frac{\partial\chi}{\partial\mu_i}\right]\frac{d\mu_i}{dt} = 0. \tag{3.4.8}$$

Equation (3.4.8) implies the following constitutive relations:

$$X_{M,i}\tau_{ij} = \rho\frac{\partial\chi}{\partial(y_{j,M})}, \quad B_i^M = \frac{\partial\chi}{\partial\mu_i}. \tag{3.4.9}$$

For rotational invariance, χ can be reduced to a function of the following inner products only [16]:

$$C_{KL} = y_{i,K}y_{i,L}, \quad N_L = y_{i,L}\mu_i. \tag{3.4.10}$$

We will use E_{KL} instead of C_{KL}:

$$E_{KL} = (C_{KL} - \delta_{KL})/2. \tag{3.4.11}$$

Therefore, we take

$$\chi = \chi(E_{KL}; N_K). \tag{3.4.12}$$

Then the constitutive relations in Eq. (3.4.9) become

$$\tau_{ij} = \rho y_{i,M}\frac{\partial\chi}{\partial E_{ML}}y_{j,L} + \rho y_{i,M}\frac{\partial\chi}{\partial N_M}\mu_j, \tag{3.4.13}$$

$$B_i^M = y_{i,L}\frac{\partial\chi}{\partial N_L}. \tag{3.4.14}$$

As an example, χ may be taken as

$$\chi = \frac{1}{2\rho^0}c_{KLMN}E_{KL}E_{MN} + \frac{1}{2}\rho^0{}_2\chi_{KL}N_KN_L$$

$$+ h_{KLM}N_KE_{LM} + \rho^0 b_{KLMN}N_KN_LE_{MN}\ldots, \tag{3.4.15}$$

where

$$c_{KLMN} - \text{second order elastic constants,}$$
$$2\chi_{KL} - \text{second order anisotropy constants,}$$
$$h_{KLM} - \text{piezomagnetic constants,}$$
$$b_{KLMN} - \text{magnetostrictive constants.}$$
(3.4.16)

For small deformations and weak magnetic fields, we have

$$J \cong 1 + u_{m,m},$$

$$\rho = \frac{\rho^0}{J} \cong \frac{\rho^0}{1 + u_{m,m}} \cong \rho^0(1 - u_{m,m}),$$
(3.4.17)

$$E_{KL} \to S_{kl} = \frac{1}{2}(u_{k,l} + u_{l,k}).$$

Since the magnetization is also small, we have, approximately,

$$M_k = \rho\mu_k \cong \rho^0(1 - u_{m,m})\mu_k \cong \rho^0\mu_k,$$
$$N_L = y_{i,L}\mu_i = (\delta_{iL} + u_{i,L})\mu_i \cong \delta_{iL}\mu_i.$$
(3.4.18)

Then the enthalpy density in Eq. (3.4.15) may be approximated by

$$\rho^0\chi \cong \frac{1}{2}c_{ijkl}S_{ij}S_{kl} + \frac{1}{2}2\chi_{kl}M_kM_l + h_{klm}M_kS_{lm}.$$
(3.4.19)

From Eqs. (3.4.13), (3.4.14) and (3.4.19), we obtain the following linear constitutive relations for piezomagnetics:

$$\tau_{ij} = c_{ijkl}S_{kl} + h_{kij}M_k,$$
$$B_i^M = 2\chi_{ik}M_k + h_{ilm}S_{lm}.$$
(3.4.20)

3.5 Thermal and Dissipative Effects

When thermal and dissipative effects are present, the integral balance laws in Eqs. (3.2.1)–(3.2.5) remain the same. However, the energy equation in Eq. (3.2.6) needs to be generalized to include thermal effects:

$$\frac{d}{dt}\int_v \rho\left(\frac{1}{2}\mathbf{v}\cdot\mathbf{v} + \varepsilon\right)dv = \int_s (\mathbf{t}\cdot\mathbf{v} - \mathbf{n}\cdot\mathbf{q})\,ds$$

$$+ \int_v \left[\mathbf{f}\cdot\mathbf{v} - \rho\boldsymbol{\mu}\cdot\frac{\partial\mathbf{B}^M}{\partial t} + \rho r\right]dv. \quad (3.5.1)$$

In addition, the second law of thermodynamics needs to be added:

$$\frac{d}{dt}\int_v \rho\eta\,dv \geq \int_v \frac{\rho r}{\theta}\,dv - \int_s \frac{\mathbf{q}\cdot\mathbf{n}}{\theta}\,ds. \quad (3.5.2)$$

Equations (3.5.1) and (3.5.2) can be converted into differential forms as

$$\rho \frac{d\varepsilon}{dt} = \tau_{ij} v_{j,i} - f_j^M v_j - \rho \mu_j \frac{\partial B_j^M}{\partial t} + \rho r - q_{i,i}, \tag{3.5.3}$$

$$\rho \frac{d\eta}{dt} \geq \frac{\rho r}{\theta} - \left(\frac{q_i}{\theta}\right)_{,i}. \tag{3.5.4}$$

The energy equation in Eq. (3.5.3) can be further written as

$$\rho \frac{d\varepsilon}{dt} = \tau_{ij} v_{j,i} - \rho \mu_j \frac{dB_j^M}{dt} + \rho r - q_{i,i}. \tag{3.5.5}$$

We introduce

$$\chi = \varepsilon + B_i^M \mu_i. \tag{3.5.6}$$

Then the energy equation in Eq. (3.5.5) takes the following form:

$$\rho \frac{d\chi}{dt} = \tau_{ij} v_{j,i} + B_j^M \rho \frac{d\mu_j}{dt} + \rho r - q_{i,i}. \tag{3.5.7}$$

Eliminating r from Eqs. (3.5.4) and (3.5.7), we obtain the Clausius–Duhem inequality as

$$\rho \left(\theta \frac{d\eta}{dt} - \frac{d\chi}{dt}\right) + \tau_{ij} v_{j,i} + \rho B_j^M \frac{d\mu_j}{dt} - \frac{q_i}{\theta} \theta_{,i} \geq 0. \tag{3.5.8}$$

Under the Legendre transform,

$$F = \chi - \theta\eta, \quad \chi = F + \theta\eta, \tag{3.5.9}$$

the Clausius–Duhem inequality in Eq. (3.5.8) becomes

$$-\rho \left(\frac{dF}{dt} + \frac{d\theta}{dt}\eta\right) + \tau_{ij} v_{j,i} + \rho B_j^M \frac{d\mu_j}{dt} - \frac{q_i}{\theta} \theta_{,i} \geq 0. \tag{3.5.10}$$

At the same time, the energy equation in Eq. (3.5.7) takes the following form:

$$\rho \left(\frac{dF}{dt} + \frac{d\theta}{dt}\eta + \theta \frac{d\eta}{dt}\right) = \tau_{ij} v_{j,i} + \rho B_j^M \frac{d\mu_j}{dt} + \rho r - q_{i,i}. \tag{3.5.11}$$

For constitutive relations, we break $\boldsymbol{\tau}$ and $\mathbf{B}^M$ into recoverable and dissipative parts through

$$\boldsymbol{\tau} = {}^R\boldsymbol{\tau} + {}^D\boldsymbol{\tau}, \quad \mathbf{B}^M = {}^R\mathbf{B}^M + {}^D\mathbf{B}^M. \tag{3.5.12}$$

The recoverable parts of Eq. (3.5.12) satisfy

$$\rho \frac{dF}{dt} = {}^R\tau_{ij} v_{j,i} + {}^R B_j^M \rho \frac{d\mu_j}{dt} - \rho\eta \frac{d\theta}{dt}. \tag{3.5.13}$$

Then the energy equation in Eq. (3.5.11) and the Clausius–Duhem inequality in Eq. (3.5.10) reduce to

$$\rho\theta\frac{d\eta}{dt} = {}^{D}\tau_{ij}v_{j,i} + {}^{D}B_{j}^{M}\rho\frac{d\mu_{j}}{dt} + \rho r - q_{i,i}, \tag{3.5.14}$$

$$ {}^{D}\tau_{ij}v_{j,i} + {}^{D}B_{j}^{M}\rho\frac{d\mu_{j}}{dt} - \frac{q_{i}}{\theta}\theta_{,i} \geq 0. \tag{3.5.15}$$

Equation (3.5.14) is the dissipation equation. With

$$v_{j,i} = X_{M,i}\frac{d}{dt}(y_{j,M}), \tag{3.5.16}$$

we write Eq. (3.5.13) as

$$\rho\frac{dF}{dt} = {}^{R}\tau_{ij}X_{M,i}\frac{d}{dt}(y_{j,M}) + {}^{R}B_{i}^{M}\rho\frac{d\mu_{i}}{dt} - \rho\eta\frac{d\theta}{dt}. \tag{3.5.17}$$

Based on Eq. (3.5.17), we let

$$F = F(y_{i,M}; \mu_{i}; \theta). \tag{3.5.18}$$

When Eq. (3.5.18) is substituted into Eq. (3.5.17), it leads to the following recoverable constitutive relations:

$$X_{M,i}{}^{R}\tau_{ij} = \rho\frac{\partial F}{\partial(y_{j,M})}, \quad {}^{R}B_{i}^{M} = \frac{\partial F}{\partial\mu_{i}}, \quad \eta = -\frac{\partial F}{\partial\theta}. \tag{3.5.19}$$

Similar to the previous section, Eq. (3.5.18) still needs to satisfy the rotational invariance [16]. The dissipative constitutive relations are restricted by Eq. (3.5.15).

Chapter 4

RIGID FERROMAGNETIC INSULATORS

The continuum theory of elastic and saturated ferromagnets in [16–19] is rather complicated. It is helpful to examine its special case of rigid, stationary and saturated ferromagnets first as a preparation or transition. Besides, rigid ferromagnets form a useful field by themselves. This chapter establishes the macroscopic theory of rigid ferromagnets. It will be shown that the well-known Landau–Lifshitz–Gilbert theory of rigid ferromagnets naturally follows from the theoretical framework of this chapter. Different from [16–19], where the magnetic field $\mathbf{H}$ is used to express the magnetic force and couple on a magnetic moment, the magnetic induction $\mathbf{B}$ is used in this chapter as explained at the beginning of Chapter 3. The SI units are used. In this chapter, the spatial coordinates are written as x_k or $\mathbf{x}$. V, S and C represent volumes, surfaces and curves fixed in space.

4.1 Saturated Ferromagnets

Consider the following field of magnetization density per unit volume:

$$\mathbf{M} = \mathbf{M}(\mathbf{x}, t). \tag{4.1.1}$$

For saturated ferromagnets, we have the following saturation condition:

$$\mathbf{M} \cdot \mathbf{M} = M_s^2, \tag{4.1.2}$$

where M_s is a constant (saturation magnetization) and s is not a tensor index. Mathematically, Eq. (4.1.2) is a constraint on $\mathbf{M}$. With differentiations with respect to t and/or $\mathbf{x}$, Eq. (4.1.2) implies that

$$M_k \frac{\partial M_k}{\partial t} = 0,$$

$$M_k M_{k,l} = 0, \tag{4.1.3}$$

$$M_{k,l} \frac{\partial M_k}{\partial t} + M_k \frac{\partial M_{k,l}}{\partial t} = 0.$$

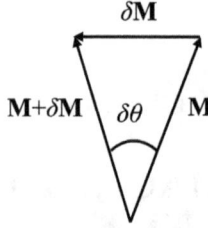

Fig. 4.1.   Change of direction of **M**.

Although the magnitude of $\mathbf{M}(\mathbf{x},t)$ cannot change because of saturation, $\mathbf{M}$ can still change its direction as described by $\delta\theta = |\delta\mathbf{M}|/|\mathbf{M}|$ in Fig. 4.1. We introduce a small rotation or angular displacement vector $\delta\boldsymbol{\theta}$ by

$$\delta\theta = \frac{|\delta\mathbf{M}|}{|\mathbf{M}|},$$

$$\delta\boldsymbol{\theta} = \delta\theta\frac{\mathbf{M}}{|\mathbf{M}|} \times \frac{\delta\mathbf{M}}{|\delta\mathbf{M}|} \tag{4.1.4}$$

$$= \frac{|\delta\mathbf{M}|}{|\mathbf{M}|}\frac{\mathbf{M}}{|\mathbf{M}|} \times \frac{\delta\mathbf{M}}{|\delta\mathbf{M}|} = \frac{1}{M_s^2}\mathbf{M} \times (\delta\mathbf{M}).$$

Then

$$\delta\boldsymbol{\theta} \times \mathbf{M} = \frac{1}{M_s^2}(\mathbf{M} \times \delta\mathbf{M}) \times \mathbf{M}$$

$$= \frac{1}{M_s^2}[(\mathbf{M} \cdot \mathbf{M})\delta\mathbf{M} - (\mathbf{M} \cdot \delta\mathbf{M})\mathbf{M}] = \delta\mathbf{M}, \tag{4.1.5}$$

where the following vector identity has been used [14]:

$$\mathbf{A} \times (\mathbf{B} \times \mathbf{C}) = (\mathbf{A} \cdot \mathbf{C})\mathbf{B} - (\mathbf{A} \cdot \mathbf{B})\mathbf{C}. \tag{4.1.6}$$

We also introduce an angular velocity vector for a saturated $\mathbf{M}$ through

$$\boldsymbol{\omega} = \lim_{\delta t \to 0}\frac{\delta\boldsymbol{\theta}}{\delta t} = \frac{1}{M_s^2}\mathbf{M} \times \frac{\partial\mathbf{M}}{\partial t}. \tag{4.1.7}$$

The power of a magnetic couple $\boldsymbol{\Gamma} = \mathbf{M} \times \mathbf{B}$ on a saturated $\mathbf{M}$ during an angular motion of $\mathbf{M}$ is given by

$$\boldsymbol{\Gamma} \cdot \boldsymbol{\omega} = (\mathbf{M} \times \mathbf{B}) \cdot \left(\frac{1}{M_s^2}\mathbf{M} \times \frac{\partial\mathbf{M}}{\partial t}\right)$$

$$= \frac{1}{M_s^2}\left[(\mathbf{M} \cdot \mathbf{M})\left(\mathbf{B} \cdot \frac{\partial\mathbf{M}}{\partial t}\right) - \left(\mathbf{M} \cdot \frac{\partial\mathbf{M}}{\partial t}\right)(\mathbf{B} \cdot \mathbf{M})\right]$$

$$= \mathbf{B} \cdot \frac{\partial\mathbf{M}}{\partial t}, \tag{4.1.8}$$

where we have used the following vector identity [14]:

$$(\mathbf{u} \times \mathbf{v}) \cdot (\mathbf{w} \times \mathbf{t}) = (\mathbf{u} \cdot \mathbf{w})(\mathbf{v} \cdot \mathbf{t}) - (\mathbf{u} \cdot \mathbf{t})(\mathbf{v} \cdot \mathbf{w}). \qquad (4.1.9)$$

Following [16], we write the angular momentum of $\mathbf{M}$ as

$$\mathbf{L} = \frac{1}{\gamma}\mathbf{M}, \qquad (4.1.10)$$

where γ is the gyromagnetic ratio which is a negative number. The evolution of $\mathbf{L}$ under a torque $\mathbf{\Gamma}$ is governed by the following angular momentum equation:

$$\frac{\partial \mathbf{L}}{\partial t} = \mathbf{\Gamma} \quad \text{or} \quad \frac{1}{\gamma}\frac{\partial \mathbf{M}}{\partial t} = \mathbf{\Gamma}. \qquad (4.1.11)$$

When $\mathbf{\Gamma} = \mathbf{M} \times \mathbf{B}$, Eq. (4.1.11) becomes

$$\frac{1}{\gamma}\frac{\partial \mathbf{M}}{\partial t} = \mathbf{M} \times \mathbf{B}. \qquad (4.1.12)$$

In this case, Eq. (4.1.8) reduces to

$$\mathbf{B} \cdot \frac{\partial \mathbf{M}}{\partial t} = \mathbf{B} \cdot \gamma(\mathbf{M} \times \mathbf{B}) = 0. \qquad (4.1.13)$$

4.2 Two-Continuum Model

Saturated ferromagnetic solids can be modeled by the two interpenetrating and interacting continua shown in Figs. 4.2 and 4.3 [16] which are more

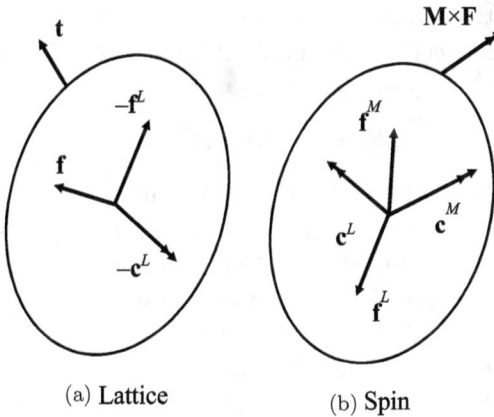

(a) Lattice   (b) Spin

Fig. 4.2. Separate lattice and spin continua: (a) lattice; (b) spin.

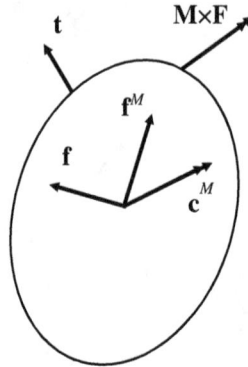

Fig. 4.3.   Combined continuum of lattice and spin.

sophisticated than Figs. 3.1 and 3.2. One of the two continua is the spin
continuum which carries distributed magnetic moments. The other is the
lattice continuum which is rigid in this chapter. The two continua cannot
displace relatively with respect to each other, but the magnetic moments
can change their directions with respect to the lattice. The two continua
interact through a local force $\mathbf{f}^L$ and a local couple $\mathbf{c}^L$ produced by an
effective local magnetic induction $\mathbf{B}^L$. We have

$$\mathbf{c}^L = \mathbf{M} \times \mathbf{B}^L, \tag{4.2.1}$$

which, as a moment vector, is shown by a double arrow in the figure.
Different from Chapter 3, the spin continuum has angular momentum in
this chapter but not linear momentum. It experiences a magnetic body
force $\mathbf{f}^M$ and a magnetic body couple $\mathbf{c}^M$ under the Maxwellian magnetic
induction $\mathbf{B}^M$. $\mathbf{c}^M$ is given by

$$\mathbf{c}^M = \mathbf{M} \times \mathbf{B}^M. \tag{4.2.2}$$

$\mathbf{f}^M$ is balanced by $\mathbf{f}^L$, i.e., $\mathbf{f}^M + \mathbf{f}^L = 0$ because the spin continuum does
not have linear momentum. Different from Chapter 3, the spin continuum
also experiences a distributed couple $\mathbf{M} \times \mathbf{F}$ per unit area on its boundary
surface in this chapter. $\mathbf{F}$ is an effective exchange field whose nature
is quantum mechanical. It describes the short range interaction among
neighboring magnetic moments. The use of a surface distribution of $\mathbf{F}$ for
the exchange interaction is the key of the formulation in [16–18]. Since
$\mathbf{F}$ and $\mathbf{B}^L$ act on $\mathbf{M}$ through cross products, the components of $\mathbf{F}$ and
$\mathbf{B}^L$ along $\mathbf{M}$ have no contributions. Without loss of generality, it can be

assumed that [16]

$$\mathbf{F} \cdot \mathbf{M} = 0 \Rightarrow \frac{\partial \mathbf{F}}{\partial t} \cdot \mathbf{M} + \mathbf{F} \cdot \frac{\partial \mathbf{M}}{\partial t} = 0, \tag{4.2.3}$$

$$\mathbf{B}^L \cdot \mathbf{M} = 0 \Rightarrow \frac{\partial \mathbf{B}^L}{\partial t} \cdot \mathbf{M} + \mathbf{B}^L \cdot \frac{\partial \mathbf{M}}{\partial t} = 0. \tag{4.2.4}$$

The lattice continuum is under the usual mechanical surface traction $\mathbf{t}$ and mechanical body force $\mathbf{f}$, in addition to the interactions with the spin continuum through $-\mathbf{f}^L$ and $-\mathbf{c}^L$. The combined continuum in Fig. 4.3 shows the loads external to the two continua only without their interactions which are internal.

4.3 Separated Spin Continuum

In this section, for the most basic understanding of the spin continuum, we treat the spin continuum in Fig. 4.2(b) alone with $\mathbf{B}^L$ and $\mathbf{B}^M$ as magnetic loads on the spin continuum. The relevant balance laws are

$$\int_S \mathbf{n} \cdot \mathbf{B}^M dS = 0, \tag{4.3.1}$$

$$\oint_C \mathbf{H} \cdot \mathbf{dl} = 0, \tag{4.3.2}$$

$$\frac{\partial}{\partial t} \int_V \frac{\mathbf{M}}{\gamma} dV = \int_S \mathbf{M} \times \mathbf{F} dS + \int_V \mathbf{M} \times (\mathbf{B}^M + \mathbf{B}^L) dV, \tag{4.3.3}$$

$$\frac{\partial}{\partial t} \int_V \rho \varepsilon dV = \int_S \mathbf{F} \cdot \frac{\partial \mathbf{M}}{\partial t} dS + \int_V -\mathbf{M} \cdot \frac{\partial}{\partial t} (\mathbf{B}^M + \mathbf{B}^L) dV, \tag{4.3.4}$$

where ρ is a constant for rigid materials. The kinetic energy associated with the angular momentum of the spin continuum has been omitted because its time derivative vanishes on account of Eq. (4.1.2). Equation (4.3.3) is the angular momentum equation. The linear momentum equation simply implies that $\mathbf{f}^M + \mathbf{f}^L = 0$. Equation (4.3.4) is the energy equation. We note that Eq. (4.1.8) has been used for the first term on the right-hand side of Eq. (4.3.4) [15, 16], and Eq. (2.9.25) has been used for the second term on the right-hand side of Eq. (4.3.4) [15]. This is the same as [15] but is different from [16], where Eq. (4.1.8) was used for both of the two terms on the right-hand side of Eq. (4.3.4). The difference between the two approaches is immaterial because of Eq. (4.2.3).

The differential forms of Eqs. (4.3.1) and (4.3.2) for the quasistatic magnetic fields are

$$\nabla \cdot \mathbf{B}^M = 0, \tag{4.3.5}$$

$$\nabla \times \mathbf{H} = 0, \quad \mathbf{H} = -\nabla \psi, \tag{4.3.6}$$

where ψ is the magnetostatic potential. We also have the following relationship:

$$\mathbf{B}^M = \mu_0(\mathbf{H} + \mathbf{M}) = \mu_0(-\nabla \psi + \mathbf{M}). \tag{4.3.7}$$

The substitution of Eq. (4.3.7) into Eq. (4.3.5) yields an equation for ψ with coupling to $\mathbf{M}$.

For the angular momentum equation in Eq. (4.3.3), we introduce an exchange tensor $\mathbf{A}$ by [16]

$$\mathbf{F} = -\mathbf{n} \cdot \mathbf{A}, \tag{4.3.8}$$

which is restricted by Eq. (4.2.3) through

$$\mathbf{A} \cdot \mathbf{M} = 0. \tag{4.3.9}$$

Then the angular momentum equation in Eq. (4.3.3) can be brought into the following differential form using Eq. (4.3.8) and the divergence theorem:

$$\varepsilon_{ijk} M_j(-A_{lk,l} + B_k^M + B_k^L) - \varepsilon_{ijk} A_{lk} M_{j,l} = \frac{1}{\gamma}\frac{\partial M_i}{\partial t}. \tag{4.3.10}$$

As a cross product, the first term on the left-hand side of Eq. (4.3.10) is perpendicular to $\mathbf{M}$. Dotting both sides of Eq. (4.3.10) by $\mathbf{M}$, we have

$$-M_i \varepsilon_{ijk} A_{lk} M_{j,l} = \frac{1}{\gamma} M_i \frac{\partial M_i}{\partial t}. \tag{4.3.11}$$

The right-hand side of Eq. (4.3.11) vanishes because of the constraint due to saturation in Eq. (4.1.3)$_1$. Then Eq. (4.3.11) reduces to

$$-M_i \varepsilon_{ijk} A_{lk} M_{j,l} = 0. \tag{4.3.12}$$

To satisfy Eq. (4.3.12), we impose the following restriction on $\mathbf{A}$ [16]:

$$A_{lk} M_{j,l} = A_{lj} M_{k,l}. \tag{4.3.13}$$

With the use of Eq. (4.3.13), the angular momentum equation in Eq. (4.3.10) reduces to

$$\frac{1}{\gamma}\frac{\partial M_i}{\partial t} = \varepsilon_{ijk} M_j(-A_{lk,l} + B_k^M + B_k^L). \tag{4.3.14}$$

Equation (4.3.14) can be written as

$$\frac{1}{\gamma}\frac{\partial \mathbf{M}}{\partial t} = \mathbf{M} \times \mathbf{B}^{eff}, \tag{4.3.15}$$

where

$$\mathbf{B}^{eff} = \mathbf{B}^{ex} + \mathbf{B}^M + \mathbf{B}^L,$$
$$B_k^{ex} = -A_{lk,l}. \tag{4.3.16}$$

Equation (4.3.15) is known as the Landau–Lifshitz equation in the literature. $\mathbf{B}^{eff}$ is the total effective magnetic induction. $\mathbf{B}^{ex}$ is the effective exchange induction. Obviously, when $\partial \mathbf{M}/\partial t = 0$, $\mathbf{M}$ assumes static equilibrium along $\mathbf{B}^{eff}$. If we take a dot product of both sides of the angular momentum equation in Eq. (4.3.14) with the following vector which is along the $\boldsymbol{\omega}$ in Eq. (4.1.7):

$$\varepsilon_{imn} M_m \frac{\partial M_n}{\partial t}, \tag{4.3.17}$$

we obtain

$$\frac{\partial M_k}{\partial t}(-A_{lk,l} + B_k^M + B_k^L) = 0, \tag{4.3.18}$$

which will be useful later.

The energy equation in Eq. (4.3.4) can be converted to differential form using Eq. (4.3.8) and the divergence theorem as

$$\rho\frac{\partial \varepsilon}{\partial t} = -A_{ij,i}\left(\frac{\partial M_j}{\partial t}\right) - A_{ij}\left(\frac{\partial M_j}{\partial t}\right)_{,i} - M_j\frac{\partial B_j^M}{\partial t} - M_j\frac{\partial B_j^L}{\partial t}. \tag{4.3.19}$$

With the use of Eq. (4.3.18), the energy equation becomes

$$\rho\frac{\partial \varepsilon}{\partial t} = -B_j^M\frac{\partial M_j}{\partial t} - B_j^L\frac{\partial M_j}{\partial t}$$
$$- A_{ij}\left(\frac{\partial M_j}{\partial t}\right)_{,i} - M_j\frac{\partial B_j^M}{\partial t} - M_j\frac{\partial B_j^L}{\partial t}$$
$$= -A_{ij}\left(\frac{\partial M_j}{\partial t}\right)_{,i} - B_j^M\frac{\partial M_j}{\partial t} - M_j\frac{\partial B_j^M}{\partial t}, \tag{4.3.20}$$

where Eq. (4.2.4) has been used. For constitutive relations, we introduce an enthalpy density χ through the following Legendre transform of ε:

$$\rho\chi = \rho\varepsilon + B_i^M M_i, \tag{4.3.21}$$

Then the energy equation in Eq. (4.3.20) reduces to

$$\rho\frac{\partial \chi}{\partial t} = -A_{ij}\left(\frac{\partial M_j}{\partial t}\right)_{,i} = -A_{ij}\frac{\partial M_{j,i}}{\partial t}. \qquad (4.3.22)$$

Equation (4.3.22) suggests that χ may be taken as

$$\chi = \chi(M_{j,i}) \qquad (4.3.23)$$

which shows a magnetization gradient theory because χ depends on the gradient of **M** instead of **M** itself. Then

$$\frac{\partial \chi}{\partial t} = \frac{\partial \chi}{\partial(M_{j,i})}\frac{\partial}{\partial t}(M_{j,i}). \qquad (4.3.24)$$

Because of Eq. (4.1.3), the nine components of $\partial(M_{j,i})/\partial t$ are not independent, and $\partial(M_{j,i})/\partial t$ are related to $\partial(M_j)/\partial t$. We substitute Eqs. (4.3.24) and (4.1.3)$_{1,3}$ into Eq. (4.3.22) using Lagrange multipliers λ and L_i [16]. This results in

$$\rho\frac{\partial \chi}{\partial(M_{j,i})}\frac{\partial}{\partial t}(M_{j,i}) = -A_{ij}\frac{\partial}{\partial t}(M_{j,i}) + \lambda M_i\frac{\partial M_i}{\partial t}$$

$$+ L_i\left[M_{j,i}\frac{\partial M_j}{\partial t} + M_j\frac{\partial}{\partial t}(M_{j,i})\right], \quad (4.3.25)$$

or

$$(\lambda M_i + L_j M_{i,j})\frac{\partial M_i}{\partial t} - \left[A_{ij} - L_i M_j + \rho\frac{\partial \chi}{\partial(M_{j,i})}\right]\frac{\partial}{\partial t}(M_{j,i}) = 0. \qquad (4.3.26)$$

Equations (4.3.26) implies that

$$\lambda M_i + L_j M_{i,j} = 0,$$

$$A_{ij} = -\rho\frac{\partial \chi}{\partial(M_{j,i})} + L_i M_j. \qquad (4.3.27)$$

Since we have assumed Eq. (4.3.9) which imposes a restriction on **A**, we can use Eq. (4.3.9) to determine L_i. From Eq. (4.3.9) which is an implication of Eq. (4.2.3), we have $\mathbf{A} \cdot \mathbf{M} = 0$. With the use of Eq. (4.3.27)$_2$, $\mathbf{A} \cdot \mathbf{M} = 0$ implies that

$$A_{ij}M_j = -\rho\frac{\partial \chi}{\partial(M_{j,i})}M_j + L_i M_j M_j = 0. \qquad (4.3.28)$$

Equation (4.3.28) determines

$$L_i = \rho \frac{1}{M_s^2} \frac{\partial \chi}{\partial(M_{k,i})} M_k. \tag{4.3.29}$$

Then, doting Eq. $(4.3.27)_1$ by $\mathbf{M}$, we have

$$\lambda M_s^2 + L_j M_i M_{i,j} = 0, \tag{4.3.30}$$

or

$$\lambda M_s^2 = -L_j M_i M_{i,j} = 0, \tag{4.3.31}$$

where Eq. $(4.1.3)_2$ has been used. Hence, the final constitutive relation for $\mathbf{A}$ is given by

$$A_{ij} = -\rho \frac{\partial \chi}{\partial(M_{j,i})} + \rho \frac{1}{M_s^2} \frac{\partial \chi}{\partial(M_{k,i})} M_k M_j. \tag{4.3.32}$$

In summary, the field equations are

$$\nabla \cdot \mathbf{B}^M = 0, \tag{4.3.33}$$

$$\mathbf{H} = -\nabla \psi, \tag{4.3.34}$$

$$\frac{1}{\gamma} \frac{\partial \mathbf{M}}{\partial t} = \mathbf{M} \times (\mathbf{B}^{ex} + \mathbf{B}^M + \mathbf{B}^L). \tag{4.3.35}$$

The constitutive relations are given by

$$\chi = \chi(M_{j,i}), \tag{4.3.36}$$

$$A_{ij} = -\rho \frac{\partial \chi}{\partial(M_{j,i})} + \rho \frac{1}{M_s^2} \frac{\partial \chi}{\partial(M_{k,i})} M_k M_j, \tag{4.3.37}$$

$$B_k^{ex} = -A_{lk,l}, \tag{4.3.38}$$

where $\mathbf{A}$ is restricted by

$$A_{lk} M_{j,l} = A_{lj} M_{k,l}. \tag{4.3.39}$$

In addition, we have

$$\mathbf{B}^M = \mu_0 (\mathbf{H} + \mathbf{M}). \tag{4.3.40}$$

$\mathbf{B}^L$ needs to be given separately. Then Eqs. (4.3.33) and (4.3.35) can be written as four equations for ψ and the three components of $\mathbf{M}$. On a boundary surface with an outward unit normal $\mathbf{n}$, possible boundary conditions are the prescriptions of [16–19]

$$\psi \quad \text{or} \quad \mathbf{n} \cdot \mathbf{B}^M, \tag{4.3.41}$$

and

$$\delta\boldsymbol{\theta} \quad \text{or} \quad \mathbf{n} \cdot \mathbf{A} \times \mathbf{M}. \tag{4.3.42}$$

$\delta\boldsymbol{\theta}$ is the angular displacement or rotation of $\mathbf{M}$ (see Eq. (4.1.4)). In dynamic problems, instead of $\delta\boldsymbol{\theta}$, its time derivative $\boldsymbol{\omega}$ may be prescribed. From Eq. (4.3.8), it can be seen that $\mathbf{n} \cdot \mathbf{A} \times \mathbf{M}$ is related to the moment on $\mathbf{M}$ by the exchange interaction through $\mathbf{F}$.

4.4 Small Magnetization Gradients

As a simple example of χ, consider the case of small magnetization gradients and take

$$\rho\chi = \frac{1}{2}\alpha_{mn}M_{k,m}M_{k,n}, \quad \alpha_{mn} = \alpha_{nm}, \tag{4.4.1}$$

which is one of the many terms in a long expression in [16]. The symmetry of α_{mn} ensures that Eq. (4.3.39) is satisfied. Then Eq. (4.3.37) generates

$$A_{ij} = -\alpha_{im}M_{j,m} + \frac{1}{M_s^2}\alpha_{im}M_{k,m}M_kM_j = -\alpha_{im}M_{j,m}, \tag{4.4.2}$$

where Eq. $(4.1.3)_2$ has been used. From Eqs. (4.3.38) and (4.4.2), we have

$$B_j^{ex} = -A_{ij,i} = \alpha_{im}M_{j,mi}. \tag{4.4.3}$$

4.5 Cubic Crystal with Initial Magnetization

In the special case of cubic crystals of class (m3m) with $\alpha_{im} = \alpha\delta_{im}$ [18, 20], we have

$$A_{ij} = -\alpha\delta_{im}M_{j,m} = -\alpha M_{j,i}. \tag{4.5.1}$$

Then the effective exchange field $\mathbf{B}^{ex}$ in Eq. (4.4.3) has the following expression:

$$B_j^{ex} = -A_{ij,i} = \alpha M_{j,ii}, \quad \mathbf{B}^{ex} = \alpha\nabla^2\mathbf{M}, \tag{4.5.2}$$

which is a common expression for $\mathbf{B}^{ex}$ (or $\mathbf{H}^{ex}$) in the literature for cubic crystals, e.g., [21], where an expression like Eq. (4.5.2) is obtained for $\mathbf{H}^{ex}$ from a discrete model of spins in cubic crystals. In the rest of this section we consider the effects of $\mathbf{B}^{ex}$ and $\mathbf{B}^M$ on $\mathbf{M}$. We set $\mathbf{B}^L = 0$ which will be shown to be true for linear cubic crystals in Section 4.9. Then the angular momentum equation in Eq. (4.3.35) takes the following form:

$$\frac{\partial \mathbf{M}}{\partial t} = \gamma \mathbf{M} \times \mathbf{B}^{eff} = \gamma \mathbf{M} \times (\mathbf{B}^{ex} + \mathbf{B}^M)$$

$$= \gamma \mathbf{M} \times (\alpha \nabla^2 \mathbf{M} + \mathbf{B}^M). \tag{4.5.3}$$

Ferromagnets are characterized by initial magnetizations. Let

$$\mathbf{M} = \mathbf{M}^0 + \mathbf{m}, \quad \mathbf{B}^M = {}^0\mathbf{B}^M + \mathbf{b}^M,$$
$$\psi = \psi^0 + \tilde{\psi}. \tag{4.5.4}$$

where the initial magnetization $\mathbf{M}^0$ is static and uniform, with a known magnitude at saturation. $\mathbf{M}^0$ is accompanied by ${}^0\mathbf{B}^M$, $\mathbf{H}^0$ and ψ^0 which are governed by

$$\nabla \cdot {}^0\mathbf{B}^M = 0, \tag{4.5.5}$$

$$0 = \gamma \mathbf{M}^0 \times (\alpha \nabla^2 \mathbf{M}^0 + {}^0\mathbf{B}^M),$$

$${}^0\mathbf{B}^M = \mu_0(\mathbf{H}^0 + \mathbf{M}^0) = \mu_0(-\nabla \psi^0 + \mathbf{M}^0). \tag{4.5.6}$$

$\mathbf{M}^0$ may align with certain directions of the crystal called easy axes while other directions are called hard axes. Under some small disturbance, $\mathbf{M}^0$ changes a little and becomes $\mathbf{M}$ which is accompanied by $\mathbf{B}^M$, $\mathbf{H}$ and ψ. $\mathbf{m}$ is a small increment of $\mathbf{M}^0$. $\mathbf{m}$ may be dynamic and is accompanied by $\mathbf{b}^M$, $\mathbf{h}^M$ and $\tilde{\psi}$. The saturation condition of $\mathbf{M}$ implies that

$$\mathbf{M} \cdot \mathbf{M} = (\mathbf{M}^0 + \mathbf{m}) \cdot (\mathbf{M}^0 + \mathbf{m})$$

$$\cong \mathbf{M}^0 \cdot \mathbf{M}^0 + 2\mathbf{M}^0 \cdot \mathbf{m} = \mathbf{M}^0 \cdot \mathbf{M}^0, \tag{4.5.7}$$

$$\Rightarrow \mathbf{M}^0 \cdot \mathbf{m} \cong 0.$$

$\mathbf{M}$, $\mathbf{B}^M$, $\mathbf{H}$ and ψ satisfy Eqs. (4.3.33)–(4.3.35) and (4.3.40). Then it can be shown that the equations for the small $\mathbf{m}$, $\mathbf{b}^M$ and $\tilde{\psi}$ are, approximately,

$$\nabla \cdot \mathbf{b}^M = 0, \tag{4.5.8}$$

$$\mathbf{b}^M = \mu_0(-\nabla \tilde{\psi} + \mathbf{m}), \tag{4.5.9}$$

$$\frac{\partial \mathbf{m}}{\partial t} = \gamma \mathbf{M} \times (\alpha \nabla^2 \mathbf{m} + \mathbf{B}^M) = \gamma \mathbf{M} \times \alpha \nabla^2 \mathbf{m} + \gamma \mathbf{M} \times \mathbf{B}^M$$

$$\cong \gamma \mathbf{M}^0 \times \alpha \nabla^2 \mathbf{m} + \gamma \mathbf{m} \times {}^0\mathbf{B}^M + \gamma \mathbf{M}^0 \times \mathbf{b}^M. \tag{4.5.10}$$

Consider the case when $\mathbf{M}^0$ and $^0\mathbf{B}^M$ are both along the x_3 axis, i.e.,

$$\mathbf{M}^0 = M^0\mathbf{i}_3, \quad {}^0\mathbf{B}^M = B^0\mathbf{i}_3. \tag{4.5.11}$$

Let

$$\begin{aligned}\mathbf{m} &= m_1\mathbf{i}_1 + m_2\mathbf{i}_2,\\ \mathbf{b}^M &= b_1\mathbf{i}_1 + b_2\mathbf{i}_2 + b_3\mathbf{i}_3.\end{aligned} \tag{4.5.12}$$

In this case, the component form of Eqs. (4.5.8)–(4.5.10) are as follows:

$$b_{1,1} + b_{2,2} + b_{3,3} = 0, \tag{4.5.13}$$

$$\begin{aligned}b_1 &= \mu_0(-\tilde{\psi}_{,1} + m_1),\\ b_2 &= \mu_0(-\tilde{\psi}_{,2} + m_2),\\ b_3 &= \mu_0(-\tilde{\psi}_{,3}),\end{aligned} \tag{4.5.14}$$

$$\begin{aligned}\frac{\partial m_1}{\partial t} &= -\alpha\gamma M^0\nabla^2 m_2 + \gamma B^0 m_2 - \gamma M^0 b_2,\\ \frac{\partial m_2}{\partial t} &= \alpha\gamma M^0\nabla^2 m_1 - \gamma B^0 m_1 + \gamma M^0 b_1.\end{aligned} \tag{4.5.15}$$

Equations (4.5.13)–(4.5.15) may be further written as

$$-\nabla^2\tilde{\psi} + m_{1,1} + m_{2,2} = 0, \tag{4.5.16}$$

$$\begin{aligned}\frac{\partial m_1}{\partial t} &= -\alpha\gamma M^0\nabla^2 m_2 + \gamma\mu_0 H^0 m_2 + \gamma M^0\mu_0\tilde{\psi}_{,2},\\ \frac{\partial m_2}{\partial t} &= \alpha\gamma M^0\nabla^2 m_1 - \gamma\mu_0 H^0 m_1 - \gamma M^0\mu_0\tilde{\psi}_{,1},\end{aligned} \tag{4.5.17}$$

which are three equations for m_1, m_2 and $\tilde{\psi}$.

Consider waves governed by Eqs. (4.5.16) and (4.5.17) with x_1 and time dependence only. We have

$$-\tilde{\psi}_{,11} + m_{1,1} = 0,$$

$$\frac{\partial m_1}{\partial t} = -\alpha\gamma M^0 m_{2,11} + \gamma\mu_0 H^0 m_2, \tag{4.5.18}$$

$$\frac{\partial m_2}{\partial t} = \alpha\gamma M^0 m_{1,11} - \gamma\mu_0 H^0 m_1 - \gamma M^0\mu_0\tilde{\psi}_{,1}.$$

Let

$$\begin{aligned}m_1 &= A\sin\xi x_1\sin\omega t, \quad m_2 = C\sin\xi x_1\cos\omega t,\\ \tilde{\psi} &= D\cos\xi x_1\sin\omega t.\end{aligned} \tag{4.5.19}$$

The substitution of Eq. (4.5.19) into Eq. (4.5.18) results in three linear algebraic equations for A, C and D:

$$\xi^2 D + \xi A = 0,$$
$$\omega A = \alpha\gamma M^0\xi^2 C + \gamma\mu_0 H^0 C, \tag{4.5.20}$$
$$-\omega C = -\alpha\gamma M^0\xi^2 A - \gamma\mu_0 H^0 A + \gamma M^0\mu_0\xi D.$$

For nontrivial solutions, the determinant of the coefficient matrix of Eq. (4.5.20) has to vanish, which yields the following dispersion relation:

$$\omega^2 = (\alpha\gamma M^0\xi^2 + \gamma\mu_0 H^0)^2 + \mu_0\gamma M^0(\alpha\gamma M^0\xi^2 + \gamma\mu_0 H^0). \tag{4.5.21}$$

When $\xi = 0$, Eq. (4.5.21) determines the following frequency:

$$\omega_c^2 = (\gamma\mu_0 H^0)^2 + \mu_0\gamma M^0(\gamma\mu_0 H^0)$$
$$= \gamma\mu_0 H^0(\gamma\mu_0 H^0 + \mu_0\gamma M^0) = \gamma^2\mu_0 H^0(^0 B^M). \tag{4.5.22}$$

Waves with x_2 and time dependence only are qualitatively the same. Waves with x_3 and time dependence only can be analyzed similarly.

When viewed as an equation mainly for $\tilde{\psi}$, Eq. (4.5.16) is elliptical and does not describe wave phenomena. We are interested in spin waves which are mainly governed by Eq. (4.5.17). For some basic understanding of spin waves through a simple and maybe somewhat crude analysis, we neglect the coupling with $\tilde{\psi}$ in Eq. (4.5.17) and obtain two approximate equations for m_1 and m_2:

$$\frac{\partial m_1}{\partial t} = -\alpha\gamma M^0\nabla^2 m_2 + \gamma\mu_0 H^0 m_2,$$
$$\frac{\partial m_2}{\partial t} = \alpha\gamma M^0\nabla^2 m_1 - \gamma\mu_0 H^0 m_1, \tag{4.5.23}$$

or

$$\frac{\partial \mathbf{m}}{\partial t} \cong \gamma\mathbf{M}^0 \times \alpha\nabla^2\mathbf{m} + \gamma\mathbf{m} \times \mu_0\mathbf{H}^0. \tag{4.5.24}$$

Equation (4.5.24) will be used in Sections 4.6–4.8 for some basic behaviors of spin waves.

4.6 Precession

Consider a uniform field of **m** which may still depend on time. In this case, Eq. (4.5.24) reduces to

$$\frac{\partial \mathbf{m}}{\partial t} \cong \gamma \mathbf{m} \times \mu_0 \mathbf{H}^0, \qquad (4.6.1)$$

whose component form is

$$\frac{\partial m_1}{\partial t} = \gamma \mu_0 H^0 m_2, \qquad \frac{\partial m_2}{\partial t} = -\gamma \mu_0 H^0 m_1. \qquad (4.6.2)$$

Equation (4.6.2) implies that

$$m_1 \frac{\partial m_1}{\partial t} + m_2 \frac{\partial m_2}{\partial t} = 0, \qquad (4.6.3)$$

or

$$\frac{\partial}{\partial t}(m_1^2 + m_2^2) = \frac{\partial}{\partial t}(\mathbf{m} \cdot \mathbf{m}) = 0. \qquad (4.6.4)$$

Equation (4.6.4) shows that **m** has a constant magnitude. From Eq. (4.5.7) we know that **m** is perpendicular to $\mathbf{M}^0$. Hence, the head of **m** describes a circle in a plane perpendicular to $\mathbf{M}^0$ or $\mu_0 \mathbf{H}^0$ as shown in Fig. 4.4. In this case, **M** sweeps through a cone around $\mathbf{M}^0$ which is a precessional motion.

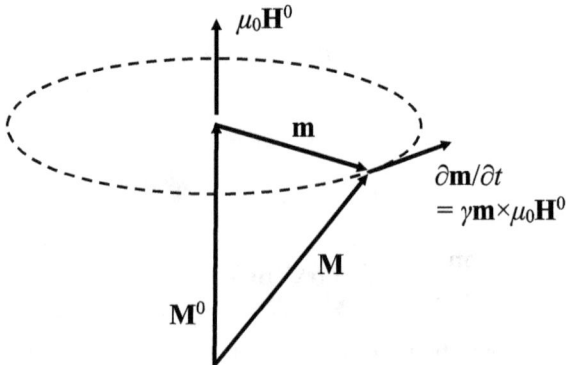

Fig. 4.4. Precessional motion of **M** around $\mu_0 \mathbf{H}^0$.

4.7   Plane Waves

Consider one-dimensional problems with x_1 and t dependence only. In this case, Eq. (4.5.23) reduces to

$$\frac{\partial m_1}{\partial t} = -\alpha\gamma M^0 \frac{\partial^2 m_2}{\partial x_1^2} + \gamma\mu_0 H^0 m_2,$$

$$\frac{\partial m_2}{\partial t} = \alpha\gamma M^0 \frac{\partial^2 m_1}{\partial x_1^2} - \gamma\mu_0 H^0 m_1. \tag{4.7.1}$$

Eliminating m_2, we obtain, from Eq. (4.7.1), a single fourth-order equation for m_1:

$$\frac{\partial^2 m_1}{\partial t^2} = -(\alpha\gamma M^0)^2 \frac{\partial^4 m_1}{\partial x_1^4}$$

$$+ 2\alpha\gamma^2 M^0 \mu_0 H^0 \frac{\partial^2 m_1}{\partial x_1^2} - (\gamma\mu_0 H^0)^2 m_1. \tag{4.7.2}$$

Consider waves propagating in the x_1 direction which is perpendicular to $\mathbf{M}^0 = M^0 \mathbf{i}_3$. Let

$$m_1 = A \exp[i(\xi x_1 - \omega t)]. \tag{4.7.3}$$

The substitution of Eq. (4.7.3) into Eq. (4.7.2) yields the following dispersion relation

$$\omega^2 = (\alpha\gamma M^0)^2 \xi^4 + 2\alpha\gamma^2 M^0 \mu_0 H^0 \xi^2 + (\gamma\mu_0 H^0)^2$$

$$= (\alpha\gamma M^0 \xi^2 + \gamma\mu_0 H^0)^2, \tag{4.7.4}$$

or

$$\omega = \pm(\alpha\gamma M^0 \xi^2 + \gamma\mu_0 H^0). \tag{4.7.5}$$

When $\xi = 0$, we have the so-called cutoff frequency ω_c below which the wave in Eq. (4.7.3) cannot propagate:

$$\omega_c = |\gamma\mu_0 H^0|. \tag{4.7.6}$$

If the second term on the right-hand side of Eq. (4.7.5) is dropped, it reduces to

$$\omega = \pm\alpha\gamma M^0 \xi^2, \tag{4.7.7}$$

which is qualitatively the same as the long wave approximation of the dispersion relation of spin waves predicted by a discrete model of spins

[22]. We note that Eq. (4.7.7) is qualitatively the same as the dispersion relation of flexural waves in elastic beams. Plane waves propagating in the direction of $\mathbf{M}^0$ can be treated similarly.

For waves with both x_1 and x_3 dependence, Eq. (4.5.23) takes the following form:

$$\frac{\partial m_1}{\partial t} = -\alpha\gamma M^0(m_{2,11} + m_{2,33}) + \gamma\mu_0 H^0 m_2,$$

$$\frac{\partial m_2}{\partial t} = \alpha\gamma M^0(m_{1,11} + m_{1,33}) - \gamma\mu_0 H^0 m_1. \qquad (4.7.8)$$

Let

$$m_1(x_1, x_3, t) = A \sin \xi x_1 \sin \zeta x_3 \sin \omega t,$$

$$m_2(x_1, x_3, t) = C \sin \xi x_1 \sin \zeta x_3 \cos \omega t. \qquad (4.7.9)$$

Substituting Eq. (4.7.9) into Eq. (4.7.8), we obtain the following equations for A and C:

$$\omega A = -\alpha\gamma M^0(-\xi^2 - \zeta^2)C + \gamma\mu_0 H^0 C,$$

$$-\omega C = \alpha\gamma M^0(-\xi^2 - \zeta^2)A - \gamma\mu_0 H^0 A. \qquad (4.7.10)$$

For nontrivial solutions, the determinant of the coefficient matrix of Eq. (4.7.10) has to vanish, which yields the following dispersion relation:

$$\begin{vmatrix} -\omega & \alpha\gamma M^0(\xi^2 + \zeta^2) + \gamma\mu_0 H^0 \\ \alpha\gamma M^0(\xi^2 + \zeta^2) + \gamma\mu_0 H^0 & -\omega \end{vmatrix} = 0, \quad (4.7.11)$$

or

$$\omega = \pm[\alpha\gamma M^0(\xi^2 + \zeta^2) + \gamma\mu_0 H^0]. \qquad (4.7.12)$$

4.8   Waves in Plates

In this section, we examine spin waves propagating in the x_1 direction of the plate in Fig. 4.5. For the so-called straight-crested waves propagating in the x_1 direction without x_2 dependence, Eq. (4.5.23) reduces to Eq. (4.7.8) which will be used in this section:

$$\frac{\partial m_1}{\partial t} = -\alpha\gamma M^0(m_{2,11} + m_{2,33}) + \gamma\mu_0 H^0 m_2,$$

$$\frac{\partial m_2}{\partial t} = \alpha\gamma M^0(m_{1,11} + m_{1,33}) - \gamma\mu_0 H^0 m_1. \qquad (4.8.1)$$

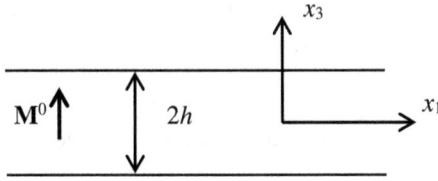

Fig. 4.5. A plate and coordinate system.

For simplicity, we consider the following boundary conditions at the top and bottom surfaces of the plate:

$$m_1 = 0, \quad m_2 = 0, \quad x_3 = \pm h. \tag{4.8.2}$$

Let

$$m_1(x_1, x_3, t) = \hat{m}_1(x_1, t) \sin \frac{n\pi x_3}{2h},$$

$$m_2(x_1, x_3, t) = \hat{m}_2(x_1, t) \sin \frac{n\pi x_3}{2h}, \tag{4.8.3}$$

$$n = 2, 4, 6, \ldots.$$

Equation (4.8.3) satisfies Eq. (4.8.2). Substituting Eq. (4.8.3) into Eq. (4.8.1), we obtain

$$\frac{\partial \hat{m}_1}{\partial t} = -\alpha\gamma M^0 \hat{m}_{2,11} + \left(\gamma\mu_0 H^0 + \alpha\gamma M^0 \frac{n^2\pi^2}{4h^2} \right) \hat{m}_2,$$

$$\frac{\partial \hat{m}_2}{\partial t} = \alpha\gamma M^0 \hat{m}_{1,11} - \left(\gamma\mu_0 H^0 + \alpha\gamma M^0 \frac{n^2\pi^2}{4h^2} \right) \hat{m}_1. \tag{4.8.4}$$

Equation (4.8.4) has the same mathematical structure as Eq. (4.7.1). Hence, the dispersion relation of Eq. (4.8.4) can be translated from Eq. (4.7.5) as

$$\omega = \pm \left(\alpha\gamma M^0 \xi^2 + \gamma\mu_0 H^0 + \alpha\gamma M^0 \frac{n^2\pi^2}{4h^2} \right). \tag{4.8.5}$$

Similarly, Eq. (4.8.5) can also be obtained from Eq. (4.7.12) by setting $\zeta = n\pi/(2h)$. The cutoff frequency of Eq. (4.8.5) is

$$\omega_c = \left| \gamma\mu_0 H^0 + \alpha\gamma M^0 \frac{n^2\pi^2}{4h^2} \right|. \tag{4.8.6}$$

If the cosine function is used in Eq. (4.8.3) for the x_3 dependence instead of the sine function, it leads to the same dispersion relation as Eq. (4.8.5) with

n assuming odd integers. Since $\gamma < 0$, we take the minus sign in Eq. (4.8.5) and normalize the dispersion relation into

$$\Omega - \Omega^0 = X^2 + n^2, \qquad (4.8.7)$$

where the dimensionless frequency Ω and dimensionless wave number X are defined by

$$\Omega = -\frac{\omega}{\alpha\gamma M^0}\frac{4h^2}{\pi^2}, \quad \Omega^0 = \frac{\mu_0 H^0}{\alpha M^0}\frac{4h^2}{\pi^2},$$

$$X = \xi\frac{2h}{\pi}. \qquad (4.8.8)$$

Equation (4.8.7) is plotted in Fig. 4.6 for $n = 1$, 2, 3 and 4. The dispersion curves in the figure represent dispersive waves. The cutoff frequencies represented by the intercepts with the vertical axis grow with n^2. This is qualitatively different from the dispersion curves of elastic waves in plates in Fig. 1.6. There are more branches of dispersion curves in the frequency range higher than what is shown in Fig. 4.6.

Next consider two unbounded plates. One is with a thickness $2h_1$. The other has a thickness $2h_2$. We assume $h_1 > h_2$. Take the minus sign in Eq. (4.8.5). In the plate of thickness $2h_1$, the dispersion relation is

$$\omega = -\alpha\gamma M^0\xi^2 - \gamma\mu_0 H^0 - \alpha\gamma M^0\frac{n^2\pi^2}{4h_1^2}. \qquad (4.8.9)$$

Similarly, for the plate of thickness $2h_2$ with the same n, the dispersion relation is

$$\omega = -\alpha\gamma M^0\xi^2 - \gamma\mu_0 H^0 - \alpha\gamma M^0\frac{n^2\pi^2}{4h_2^2}. \qquad (4.8.10)$$

$\Omega\text{-}\Omega^0$ versus X for n = 1, 2, 3, 4

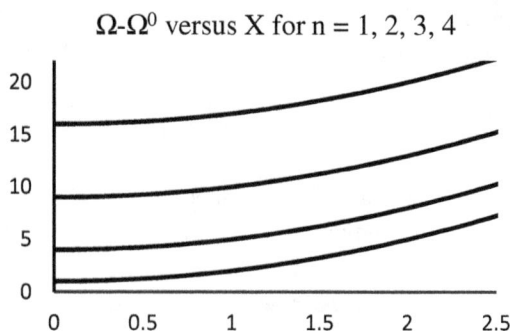

Fig. 4.6. Dispersion curves determined by Eq. (4.8.7). Abscissa $= X$. Coordinate $= \Omega - \Omega^0$.

Equations (4.8.9) and (4.8.10) may be rewritten as

$$\frac{\omega}{-\alpha\gamma M^0} = \xi^2 + \frac{\mu_0 H^0}{\alpha M^0} + \frac{n^2\pi^2}{4h_1^2}, \tag{4.8.11}$$

$$\frac{\omega}{-\alpha\gamma M^0} = \xi^2 + \frac{\mu_0 H^0}{\alpha M^0} + \frac{n^2\pi^2}{4h_2^2}, \tag{4.8.12}$$

or

$$\omega' = \xi^2 + \omega'_{c1}, \tag{4.8.13}$$

$$\omega' = \xi^2 + \omega'_{c2}, \tag{4.8.14}$$

where

$$\omega' = \frac{\omega}{-\alpha\gamma M^0},$$

$$\omega'_{c1} = \frac{\mu_0 H^0}{\alpha M^0} + \frac{n^2\pi^2}{4h_1^2} < \frac{\mu_0 H^0}{\alpha M^0} + \frac{n^2\pi^2}{4h_2^2} = \omega'_{c2}. \tag{4.8.15}$$

Consider the case when α, M^0 and H^0 are positive. From Eqs. (4.8.13) and (4.8.14),

$$\xi^2 = \omega' - \omega'_{c1}, \tag{4.8.16}$$

$$\xi^2 = \omega' - \omega'_{c2}. \tag{4.8.17}$$

Allowing ξ^2 to be negative and ξ to be purely imaginary, we plot the dispersion curves in Eqs. (4.8.16) and (4.8.17) in Fig. 4.7 qualitatively for long waves with a small wave number ξ. In the narrow frequency interval of

$$\omega'_{c1} < \omega' < \omega'_{c2}, \tag{4.8.18}$$

Eqs. (4.8.16) and (4.8.17) produce real and purely imaginary values of ξ, respectively.

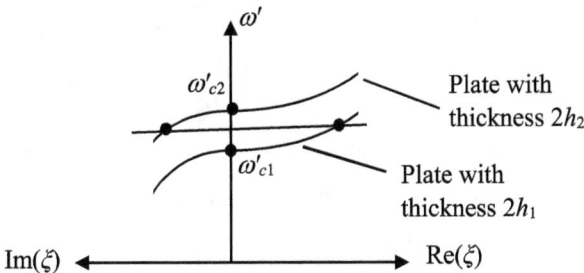

Fig. 4.7. Dispersion curves from Eqs. (4.8.16) and (4.8.17).

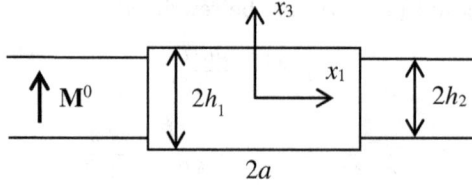

Fig. 4.8. A plate with a nonuniform thickness.

Then, in the nonuniform plate in Fig. 4.8, there may exist vibration modes that are oscillatory or sinusoidal in the central region within $|x_1| < a$ and exponentially decaying when $|x_1| > a$. Similar modes of acoustic waves in nonuniform elastic plates are called trapped modes. The existence of trapped modes and the number of trapped modes depend on the value of a. A larger a corresponds to more trapped modes.

4.9   Combined Spin-Lattice Continuum

In this section, we treat the spin and lattice continua together as shown in Fig. 4.3. In this case, the effective local magnetic induction $\mathbf{B}^L$ is an internal interaction between the two continua. Among Eqs. (4.3.1)–(4.3.4), the energy equation in Eq. (4.3.4) for the spin continuum alone needs to be replaced by the following energy equation for the combined lattice and spin continuum:

$$\frac{\partial}{\partial t}\int_V \rho\varepsilon dV = \int_S \mathbf{F}\cdot\frac{\partial \mathbf{M}}{\partial t}dS + \int_V -\mathbf{M}\cdot\frac{\partial \mathbf{B}^M}{\partial t}dV. \qquad (4.9.1)$$

Its differential form is

$$\rho\frac{\partial\varepsilon}{\partial t} = -A_{ij,i}\left(\frac{\partial M_j}{\partial t}\right) - A_{ij}\left(\frac{\partial M_j}{\partial t}\right)_{,i} - M_j\frac{\partial B_j^M}{\partial t}. \qquad (4.9.2)$$

With the use of Eq. (4.3.18), the energy equation in Eq. (4.9.2) becomes

$$\rho\frac{\partial\varepsilon}{\partial t} = -B_j^M\frac{\partial M_j}{\partial t} - B_j^L\frac{\partial M_j}{\partial t} - A_{ij}\left(\frac{\partial M_j}{\partial t}\right)_{,i} - M_j\frac{\partial B_j^M}{\partial t}. \qquad (4.9.3)$$

We introduce an enthalpy density function χ per unit volume through

$$\rho\chi = \rho\varepsilon + B_i^M M_i. \qquad (4.9.4)$$

Then the energy equation becomes

$$\rho\frac{\partial\chi}{\partial t} = -A_{ij}\left(\frac{\partial M_j}{\partial t}\right)_{,i} - B_j^L\frac{\partial M_j}{\partial t}. \qquad (4.9.5)$$

For constitutive relations, based on Eq. (4.9.5), we consider

$$\chi = \chi(M_i; M_{j,i}). \tag{4.9.6}$$

Then

$$\frac{\partial \chi}{\partial t} = \frac{\partial \chi}{\partial M_i} \frac{\partial M_i}{\partial t} + \frac{\partial \chi}{\partial (M_{j,i})} \frac{\partial}{\partial t}(M_{j,i}). \tag{4.9.7}$$

Substituting Eq. (4.9.7) into Eq. (4.9.5), including the constraints in Eq. (4.1.3)$_{1,3}$ by Lagrange multipliers λ and L_i, we obtain

$$\rho \frac{\partial \chi}{\partial M_i} \frac{\partial M_i}{\partial t} + \rho \frac{\partial \chi}{\partial (M_{j,i})} \frac{\partial}{\partial t}(M_{j,i}) = -A_{ij} \frac{\partial}{\partial t}(M_{j,i}) - B_j^L \frac{\partial M_j}{\partial t} + \lambda M_i \frac{\partial M_i}{\partial t}$$

$$+ L_i \left[M_{j,i} \frac{\partial M_j}{\partial t} + M_j \frac{\partial}{\partial t}(M_{j,i}) \right], \tag{4.9.8}$$

or

$$- \left[B_i^L - \lambda M_i - L_j M_{i,j} + \rho \frac{\partial \chi}{\partial M_i} \right] \frac{\partial M_i}{\partial t}$$

$$- \left[A_{ij} - L_i M_j + \rho \frac{\partial \chi}{\partial (M_{j,i})} \right] \frac{\partial}{\partial t}(M_{j,i}) = 0. \tag{4.9.9}$$

Equations (4.9.9) implies that

$$B_i^L = -\rho \frac{\partial \chi}{\partial M_i} + \lambda M_i + L_j M_{i,j},$$

$$A_{ij} = -\rho \frac{\partial \chi}{\partial (M_{j,i})} + L_i M_j. \tag{4.9.10}$$

Since both the lattice and spin continua are considered, a constitutive relation for $\mathbf{B}^L$ has been obtained in Eq. (4.9.10). This is fundamentally different from Section 4.3. Since we have assumed Eqs. (4.2.3) and (4.2.4) which impose restrictions on $\mathbf{F}$ and $\mathbf{B}^L$, we can use Eqs. (4.2.3) and (4.2.4) to determine L_i and λ. From Eq. (4.3.9) which is an implication of Eq. (4.2.3), we have $\mathbf{A} \cdot \mathbf{M} = 0$ or, with the use of Eq. (4.9.10)$_2$,

$$A_{ij} M_j = -\rho \frac{\partial \chi}{\partial (M_{j,i})} M_j + L_i M_j M_j = 0. \tag{4.9.11}$$

Equation (4.9.11) determines

$$L_i = \rho \frac{1}{M_s^2} \frac{\partial \chi}{\partial (M_{k,i})} M_k. \tag{4.9.12}$$

From Eq. (4.2.4), we have $\mathbf{B}^L \cdot \mathbf{M} = 0$ or, with the use of Eq. (4.9.10)$_1$,

$$B_i^L M_i = -\rho \frac{\partial \chi}{\partial M_i} M_i + \lambda M_i M_i + L_j M_i M_{i,j}$$

$$= -\rho \frac{\partial \chi}{\partial M_i} M_i + \lambda M_i M_i = 0, \qquad (4.9.13)$$

where Eq. (4.1.3)$_2$ has been used. Equation (4.9.13) determines

$$\lambda = \rho \frac{1}{M_s^2} \frac{\partial \chi}{\partial M_k} M_k. \qquad (4.9.14)$$

Using Eqs. (4.9.12) and (4.9.14), we write Eq. (4.9.10) as

$$B_i^L = -\rho \frac{\partial \chi}{\partial M_i} + \rho \frac{1}{M_s^2} \frac{\partial \chi}{\partial M_k} M_k M_i + \rho \frac{1}{M_s^2} \frac{\partial \chi}{\partial (M_{k,j})} M_k M_{i,j},$$

$$A_{ij} = -\rho \frac{\partial \chi}{\partial (M_{j,i})} + \rho \frac{1}{M_s^2} \frac{\partial \chi}{\partial (M_{k,i})} M_k M_j. \qquad (4.9.15)$$

Equation (4.9.15)$_2$ is still restricted by Eq. (4.3.13).

As an example, for small $\mathbf{M}$ with small gradients, χ may be taken as

$$\rho \chi = \frac{1}{2} \chi_{mn} M_m M_n + \frac{1}{2} \alpha_{mn} M_{k,m} M_{k,n}, \qquad (4.9.16)$$

$$\chi_{mn} = \chi_{nm}, \quad \alpha_{mn} = \alpha_{nm}.$$

Then Eq. (4.9.15) generates

$$B_i^L = -\chi_{ik} M_k + \frac{1}{M_s^2} (\chi_{mn} M_m M_n) M_i, \qquad (4.9.17)$$

$$A_{ij} = -\alpha_{im} M_{j,m}.$$

In the special case of cubic crystals with

$$\chi_{im} = \chi^M \delta_{im}, \quad \alpha_{im} = \alpha \delta_{im}, \qquad (4.9.18)$$

Eq. (4.9.17) reduces to

$$B_i^L = 0, \quad A_{ij} = -\alpha M_{j,i}. \qquad (4.9.19)$$

4.10 Thermal and Dissipative Effects

The equations in [16] include thermal and dissipative effects which were dropped in the previous sections of this chapter for simplicity. In this section, thermal and dissipative effects in rigid ferromagnets [23] are treated for completeness. The balance laws in Eqs. (4.3.1)–(4.3.3) remain the same. The energy equation in Eq. (4.9.1) for the two continua together needs to be generalized to include thermal effects as

$$\frac{\partial}{\partial t}\int_V \rho\varepsilon dV = \int_S \left(\mathbf{F}\cdot\frac{\partial \mathbf{M}}{\partial t} - \mathbf{n}\cdot\mathbf{q}\right)dS + \int_V \left(-\mathbf{M}\cdot\frac{\partial \mathbf{B}^M}{\partial t} + \rho r\right)dV,$$
(4.10.1)

where r is the body heat source per unit mass and $\mathbf{q}$ is the heat flux vector. In addition, the second law of thermal dynamics needs to be added:

$$\frac{\partial}{\partial t}\int_V \rho\eta dV \geq \int_V \frac{\rho r}{\theta}dV - \int_S \frac{\mathbf{q}\cdot\mathbf{n}}{\theta}dS,$$
(4.10.2)

where θ is the absolute temperature and η is the entropy density per unit mass. Equations (4.10.1) and (4.10.2) can be converted to differential forms as

$$\rho\frac{\partial\varepsilon}{\partial t} = -A_{ij,i}\left(\frac{\partial M_j}{\partial t}\right) - A_{ij}\left(\frac{\partial M_j}{\partial t}\right)_{,i} - M_j\frac{\partial B_j^M}{\partial t} + \rho r - q_{i,i},$$
(4.10.3)

$$\rho\frac{\partial\eta}{\partial t} \geq \frac{\rho r}{\theta} - \left(\frac{q_i}{\theta}\right)_{,i}.$$
(4.10.4)

With the use of Eq. (4.3.18), the energy equation in Eq. (4.10.3) becomes

$$\rho\frac{\partial\varepsilon}{\partial t} = -B_j^M\frac{\partial M_j}{\partial t} - B_j^L\frac{\partial M_j}{\partial t}$$
$$- A_{ij}\left(\frac{\partial M_j}{\partial t}\right)_{,i} - M_j\frac{\partial B_j^M}{\partial t} + \rho r - q_{i,i}.$$
(4.10.5)

We introduce an enthalpy function χ through

$$\rho\chi = \rho\varepsilon + B_i^M M_i.$$
(4.10.6)

Then Eq. (4.10.5) takes the following form:

$$\rho\frac{\partial\chi}{\partial t} = -A_{ij}\left(\frac{\partial M_j}{\partial t}\right)_{,i} - B_j^L\frac{\partial M_j}{\partial t} + \rho r - q_{i,i}.$$
(4.10.7)

Equation (4.10.4) can be rewritten as

$$\rho r \leq \theta\left(\frac{q_i}{\theta}\right)_{,i} + \theta\rho\frac{\partial\eta}{\partial t}$$
$$= \theta\left(\frac{q_{i,i}}{\theta} - \frac{q_i}{\theta^2}\theta_{,i}\right) + \theta\rho\frac{\partial\eta}{\partial t} = q_{i,i} - \frac{q_i}{\theta}\theta_{,i} + \theta\rho\frac{\partial\eta}{\partial t}.$$
(4.10.8)

Eliminating ρr from Eqs. (4.10.7) and (4.10.8), we obtain the Clausius–Duhem inequality as

$$\rho\frac{\partial\chi}{\partial t} + A_{ij}\left(\frac{\partial M_j}{\partial t}\right)_{,i} + B_j^L\frac{\partial M_j}{\partial t} + q_{i,i} \le q_{i,i} - \frac{q_i}{\theta}\theta_{,i} + \theta\rho\frac{\partial\eta}{\partial t}, \quad (4.10.9)$$

or

$$\theta\rho\frac{\partial\eta}{\partial t} - \rho\frac{\partial\chi}{\partial t} - A_{ij}\left(\frac{\partial M_j}{\partial t}\right)_{,i} - B_j^L\frac{\partial M_j}{\partial t} - \frac{q_i}{\theta}\theta_{,i} \ge 0. \quad (4.10.10)$$

Under the following Legendre transform:

$$F = \chi - \theta\eta, \quad (4.10.11)$$

and hence

$$\frac{\partial\chi}{\partial t} = \frac{\partial F}{\partial t} + \frac{\partial\theta}{\partial t}\eta + \theta\frac{\partial\eta}{\partial t}, \quad (4.10.12)$$

the Clausius–Duhem inequality in Eq. (4.10.10) takes the following form:

$$-\rho\left(\frac{\partial F}{\partial t} + \frac{\partial\theta}{\partial t}\eta\right) - A_{ij}\left(\frac{\partial M_j}{\partial t}\right)_{,i} - B_j^L\frac{\partial M_j}{\partial t} - \frac{q_i}{\theta}\theta_{,i} \ge 0. \quad (4.10.13)$$

At the same time the energy equation in Eq. (4.10.7) becomes

$$\rho\left(\frac{\partial F}{\partial t} + \frac{\partial\theta}{\partial t}\eta + \theta\frac{\partial\eta}{\partial t}\right) = -A_{ij}\left(\frac{\partial M_j}{\partial t}\right)_{,i} - B_j^L\frac{\partial M_j}{\partial t} + \rho r - q_{i,i}. \quad (4.10.14)$$

For constitutive relations we break $\mathbf{B}^L$ into recoverable and dissipative parts and assume

$$\mathbf{B}^L = {}^R\mathbf{B}^L(M_i; M_{j,i}; \theta) + {}^D\mathbf{B}^L(M_i; M_{j,i}; \theta; \theta_{,j}; \dot{M}_i),$$
$$\mathbf{q} = \mathbf{q}(M_i; M_{j,i}; \theta; \theta_{,j}; \dot{M}_i). \quad (4.10.15)$$

The recoverable part of Eq. (4.10.15)$_1$ satisfies

$$\rho\frac{\partial F}{\partial t} = -{}^R B_j^L\frac{\partial M_j}{\partial t} - A_{ij}\left(\frac{\partial M_j}{\partial t}\right)_{,i} - \rho\eta\frac{\partial\theta}{\partial t}. \quad (4.10.16)$$

Then the energy equation in Eq. (4.10.14) and the Clausius–Duhem inequality in Eq. (4.10.13) reduce to

$$\theta\rho\frac{\partial\eta}{\partial t} = -{}^D B_j^L\frac{\partial M_j}{\partial t} + \rho r - q_{i,i}, \quad (4.10.17)$$

$$-{}^D B_j^L\frac{\partial M_j}{\partial t} - \frac{q_i}{\theta}\theta_{,i} \ge 0. \quad (4.10.18)$$

Equation (4.10.17) is the dissipation equation. With

$$\left(\frac{\partial M_j}{\partial t}\right)_{,i} = \frac{\partial}{\partial t}(M_{j,i}),$$ (4.10.19)

we write Eq. (4.10.16) as

$$\rho\frac{\partial F}{\partial t} = -{}^R B_j^L \frac{\partial M_j}{\partial t} - A_{ij}\frac{\partial}{\partial t}(M_{j,i}) - \rho\eta\frac{\partial\theta}{\partial t}.$$ (4.10.20)

Let

$$F = F(M_i; M_{j,i}; \theta).$$ (4.10.21)

We substitute Eq. (4.10.21) into Eq. (4.10.20) and use Lagrange multipliers to introduce the constrains in Eqs. (4.1.3)$_{1,3}$. In a way similar to Section 4.9, it can be shown that the recoverable constitutive relations are given by

$$^R B_i^L = -\rho\frac{\partial F}{\partial M_i} + \lambda M_i + L_m M_{i,m},$$ (4.10.22)

$$A_{ij} = -\rho\frac{\partial F}{\partial(M_{j,i})} + L_i M_j,$$

$$\eta = -\frac{\partial F}{\partial\theta}.$$ (4.10.23)

λ and L_i can be determined in a way similar to the previous section. $^D\mathbf{B}^L$ and $\mathbf{q}$ in the constitutive relations in Eq. (4.10.15) are restricted by the Clausius–Duhem inequality in Eq. (4.10.18).

When dissipation is present, the Landau–Lifshitz equation in Eq. (4.3.35) becomes the more general Landau–Lifshitz–Gilbert equation in the literature [24]:

$$\frac{1}{\gamma}\frac{\partial\mathbf{M}}{\partial t} = \mathbf{M}\times\mathbf{B}^{eff} - \beta\mathbf{M}\times\frac{\partial\mathbf{M}}{\partial t},$$ (4.10.24)

where β is a damping coefficient. Equation (4.10.24) can be written as

$$\frac{1}{\gamma}\frac{\partial\mathbf{M}}{\partial t} = \mathbf{M}\times(\mathbf{B}^{eff} - \beta\frac{\partial\mathbf{M}}{\partial t}).$$ (4.10.25)

On the other hand, with the theoretical framework for rigid ferromagnets developed in this chapter, from Eqs. (4.3.35) and (4.10.15), we can write

$$\frac{1}{\gamma}\frac{\partial\mathbf{M}}{\partial t} = \mathbf{M}\times(\mathbf{B}^{ex} + \mathbf{B}^M + {}^R\mathbf{B}^L + {}^D\mathbf{B}^L)$$

$$= \mathbf{M}\times(\mathbf{B}^{eff} + {}^D\mathbf{B}^L).$$ (4.10.26)

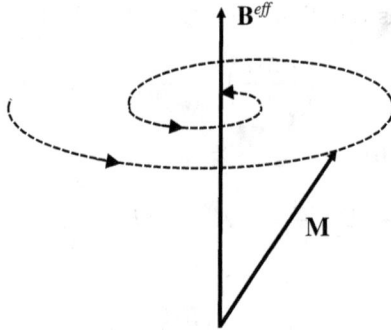

Fig. 4.9. Precessional motion of **M** with damping.

Hence, we identify

$$^D\mathbf{B}^L = -\beta\frac{\partial\mathbf{M}}{\partial t}. \tag{4.10.27}$$

Thus, the theoretical framework in this chapter can reduce to the Landau–Lifshitz–Gilbert equation. To ensure that Eq. (4.10.18) is satisfied, we impose that

$$-^D\mathbf{B}^L \cdot \frac{\partial\mathbf{M}}{\partial t} \geq 0, \quad -q_k\theta_{,k} \geq 0. \tag{4.10.28}$$

Then, with the use of Eq. (4.10.27), we have

$$\beta\frac{\partial\mathbf{M}}{\partial t} \cdot \frac{\partial\mathbf{M}}{\partial t} \geq 0, \tag{4.10.29}$$

which implies that $\beta \geq 0$. When damping is present, the precessional motion of **M** is shown in Fig. 4.9 qualitatively.

More general equations for rigid and saturated ferromagnetic conductors including thermal and electromagnetic couplings as well as dissipative effects can be found in [25].

Chapter 5

ELASTIC FERROMAGNETIC
INSULATORS

This chapter is a generalization of the previous chapter from rigid to elastic ferromagnets with a saturated magnetization. The theoretical framework follows the two-continuum model in [16–19]. The magnetic force and couple on the spin continuum are in terms of the magnetic induction $\mathbf{B}$ instead of the magnetic field $\mathbf{H}$ used in [16–19]. This is consistent with the previous chapter and the later paper [15] by the author of [16–19]. Continuum theories of saturated and elastic ferromagnets have also been developed by other authors [26–29] after [16] with similar results. This chapter is in SI units.

5.1 Two-Continuum Model

The two-continuum model in Figs. 5.1 and 5.2 for an elastic ferromagnet is similar to the two-continuum model for a rigid ferromagnet in Section 4.2. The difference is that the lattice continuum is deformable in this chapter. We follow the two-point Cartesian tensor notation of Chapter 1 with $\mathbf{X}$ and $\mathbf{y}$ for the reference and present coordinates of a material point of the lattice continuum. We still have the following expressions:

$$\mathbf{f}^M = \mathbf{M} \cdot (\mathbf{B}^M \nabla), \quad f_i^M = M_k B_{k,i}^M,$$
$$f_l^M = T_{ml,m}^M, \tag{5.1.1}$$
$$T_{ml}^M = B_m H_l + \frac{\mu_0}{2}(M_k M_k - H_k H_k)\delta_{ml},$$

$$\mathbf{c}^M = \mathbf{M} \times \mathbf{B}^M, \quad c_i^M = \varepsilon_{ijk} M_j B_k^M,$$
$$\mathbf{c}^L = \mathbf{M} \times \mathbf{B}^L, \quad c_i^L = \varepsilon_{ijk} M_j B_k^L. \tag{5.1.2}$$

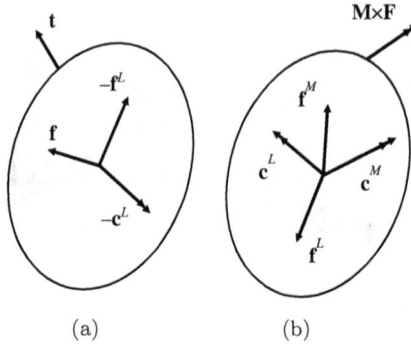

Fig. 5.1. Separate lattice and spin continua: (a) lattice; (b) spin.

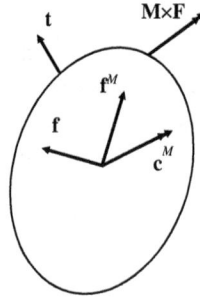

Fig. 5.2. Combined continuum of lattice and spin.

We still assume that $\mathbf{F}$ and $\mathbf{B}^L$ satisfy

$$\mathbf{F} \cdot \mathbf{M} = 0, \qquad (5.1.3)$$

$$\mathbf{B}^L \cdot \mathbf{M} = 0. \qquad (5.1.4)$$

For elastic ferromagnets, we will use the magnetization per unit mass, $\boldsymbol{\mu}$, more often than $\mathbf{M}$, the magnetization per unit volume. The saturation condition is

$$\boldsymbol{\mu} \cdot \boldsymbol{\mu} = \mu_s^2, \qquad (5.1.5)$$

where

$$\boldsymbol{\mu} = \frac{\mathbf{M}}{\rho}. \qquad (5.1.6)$$

Equation (5.1.5) implies that

$$\mu_k \frac{d\mu_k}{dt} = 0,$$

$$\mu_k \mu_{k,L} = 0, \tag{5.1.7}$$

$$\mu_{k,L} \frac{d\mu_k}{dt} + \mu_k \frac{d\mu_{k,L}}{dt} = 0.$$

Corresponding to Eq. (4.1.8), for the power of a magnetic couple $\mathbf{\Gamma} = \mathbf{M} \times \mathbf{B}$ on a saturated magnetization, we have, in terms of $\boldsymbol{\mu}$,

$$\mathbf{\Gamma} \cdot \boldsymbol{\omega} = (\rho \boldsymbol{\mu} \times \mathbf{B}) \cdot \left(\frac{1}{\mu_s^2} \boldsymbol{\mu} \times \frac{d\boldsymbol{\mu}}{dt} \right)$$

$$= \frac{\rho}{\mu_s^2} \left[(\boldsymbol{\mu} \cdot \boldsymbol{\mu}) \left(\mathbf{B} \cdot \frac{d\boldsymbol{\mu}}{dt} \right) - \left(\boldsymbol{\mu} \cdot \frac{d\boldsymbol{\mu}}{dt} \right) (\mathbf{B} \cdot \boldsymbol{\mu}) \right] = \rho \mathbf{B} \cdot \frac{d\boldsymbol{\mu}}{dt}. \tag{5.1.8}$$

From Eq. (3.1.6), we have

$$w^M = -\rho \mu_j \frac{\partial B_j^M}{\partial t} = -\rho \mu_j \frac{\partial B_j^M}{\partial t} - \rho \mu_j v_i B_{j,i}^M + \rho \mu_j v_i B_{j,i}^M$$

$$= -\rho \mu_j \frac{d B_j^M}{dt} + f_i^M v_i = -\rho \frac{d}{dt}(\mu_j B_j^M) + \rho B_j^M \frac{d\mu_j}{dt} + f_i^M v_i. \tag{5.1.9}$$

5.2 Integral Balance Laws

For the separate lattice and spin continua in Fig. 5.1 or their combination in Fig. 5.2, the relevant balance laws in integral form are

$$\int_s \mathbf{n} \cdot \mathbf{B}^M ds = 0, \tag{5.2.1}$$

$$\int_l \mathbf{H} \cdot \mathbf{dl} = 0, \tag{5.2.2}$$

$$\frac{d}{dt} \int_v \rho dv = 0, \tag{5.2.3}$$

$$\frac{d}{dt} \int_v \rho \mathbf{v} dv = \int_s \mathbf{t} ds + \int_v (\rho \mathbf{f} + \mathbf{f}^M) dv, \tag{5.2.4}$$

$$\frac{d}{dt}\int_v \mathbf{y} \times \rho\mathbf{v}\,dv = \int_s \mathbf{y} \times \mathbf{t}\,ds + \int_v \left[\mathbf{y} \times (\rho\mathbf{f} - \mathbf{f}^L) - \mathbf{M} \times \mathbf{B}^L\right]dv,$$

$$(5.2.5)$$

$$\frac{d}{dt}\int_v \frac{\mathbf{M}}{\gamma}\,dv = \int_s \mathbf{M} \times \mathbf{F}\,ds + \int_v \mathbf{M} \times (\mathbf{B}^M + \mathbf{B}^L)\,dv, \qquad (5.2.6)$$

$$\frac{d}{dt}\int_v \rho\left(\frac{1}{2}\mathbf{v} \cdot \mathbf{v} + \varepsilon\right)dv = \int_s \left(\mathbf{t} \cdot \mathbf{v} + \mathbf{F} \cdot \rho\frac{d\boldsymbol{\mu}}{dt}\right)ds$$

$$+ \int_v \left(\rho\mathbf{f} \cdot \mathbf{v} - \mathbf{M} \cdot \frac{\partial \mathbf{B}^M}{\partial t}\right)dv. \quad (5.2.7)$$

The angular momentum equations in Eqs. (5.2.5) and (5.2.6) are for the lattice and spin continua separately. The linear momentum equation in Eq. (5.2.4) and the energy equation in Eq. (5.2.7) are for the combined continuum. Since the spin continuum is assumed to be without linear momentum, the linear momentum equation for the spin continuum alone simply implies that $\mathbf{f}^L = -\mathbf{f}^M$. In the energy equation in Eq. (5.2.7), the magnetic power on s is from Eq. (5.1.8) and the magnetic power in v is given by Eq. (5.1.9). The kinetic energy associated with the angular momentum of the spin continuum has been omitted because its material time derivative vanishes [16].

5.3   Differential Balance Laws

The differential forms of Eqs. (5.2.1) and (5.2.2) are the same as those in Chapter 4:

$$\nabla \cdot \mathbf{B}^M = 0, \quad B^M_{k,k} = 0, \tag{5.3.1}$$

$$\nabla \times \mathbf{H} = 0, \quad \varepsilon_{ijk}H_{k,j} = 0. \tag{5.3.2}$$

Equation (5.3.2) allows the introduction of a scalar potential ψ so that $\mathbf{H} = -\nabla\psi$. The differential form of the conservation of mass in Eq. (5.2.3) is the same as before:

$$\rho^0 = \rho J. \tag{5.3.3}$$

With the introduction of the Cauchy stress tensor $\boldsymbol{\tau}$ such that the surface traction

$$\mathbf{t} = \mathbf{n} \cdot \boldsymbol{\tau}, \tag{5.3.4}$$

the differential form of the linear momentum equation in Eq. (5.2.4) is

$$\tau_{ij,i} + \rho f_j + f_j^M = \rho \ddot{u}_j. \tag{5.3.5}$$

In component form, the angular momentum equation of the lattice continuum in Eq. (5.2.5) takes the following form:

$$\frac{d}{dt} \int_v \varepsilon_{ijk} y_j \rho v_k dv = \int_s \varepsilon_{ijk} y_j t_k ds + \int_v [(\varepsilon_{ijk} y_j (\rho f_k + f_k^M) - c_i^L] dv. \tag{5.3.6}$$

With the use of τ and the divergence theorem, Eq. (5.3.6) becomes

$$\int_v \varepsilon_{ijk} y_j (\rho \dot{v}_k - \rho f_k - f_k^M - \tau_{lk,l}) dv = \int_v (\varepsilon_{ijk} \tau_{jk} - c_i^L) dv. \tag{5.3.7}$$

The left-hand side of Eq. (5.3.7) vanishes because of the linear momentum equation in Eq. (5.3.5). Hence,

$$\int_v (\varepsilon_{ijk} \tau_{jk} - c_i^L) dv = 0, \tag{5.3.8}$$

which implies that

$$\varepsilon_{ijk} \tau_{jk} - c_i^L = 0. \tag{5.3.9}$$

We introduce an exchange tensor $\mathbf{A}$ through [16]

$$\mathbf{F} = -\mathbf{n} \cdot \mathbf{A}. \tag{5.3.10}$$

Equation (5.1.3) implies that

$$\mathbf{A} \cdot \mathbf{M} = 0. \tag{5.3.11}$$

We also impose the following restriction on $\mathbf{A}$ (see Eq. (4.3.13) or [16]):

$$A_{lk}\mu_{j,l} = A_{lj}\mu_{k,l}. \tag{5.3.12}$$

With the use of $\mathbf{A}$ and the divergence theorem, we obtain the differential form of the angular momentum equation of the spin continuum in Eq. (5.2.6) as

$$\varepsilon_{ijk}\rho\mu_j \left(-A_{lk,l} - \frac{A_{lk}}{\rho}\rho_{,l} + B_k^M + B_k^L \right) - \varepsilon_{ijk}\rho A_{lk}\mu_{j,l} = \frac{1}{\gamma}\rho\frac{d\mu_i}{dt}. \tag{5.3.13}$$

Using Eq. (5.3.12), we reduce Eq. (5.3.13) to

$$\varepsilon_{ijk}\mu_j\left(-A_{lk,l} - \frac{A_{lk}}{\rho}\rho_{,l} + B_k^M + B_k^L\right) = \frac{1}{\gamma}\frac{d\mu_i}{dt}. \qquad (5.3.14)$$

Dotting both sides of Eq. (5.3.14) by

$$\varepsilon_{imn}\mu_m\frac{d\mu_n}{dt}, \qquad (5.3.15)$$

we obtain

$$\frac{d\mu_k}{dt}\left(-A_{lk,l} - \frac{A_{lk}}{\rho}\rho_{,l} + B_k^M + B_k^L\right) = 0, \qquad (5.3.16)$$

which will be useful in the following. The energy equation in Eq. (5.2.7) can be brought into differential form as

$$\rho\frac{d\varepsilon}{dt} = \tau_{ij}v_{j,i} - A_{ij,i}\rho\left(\frac{d\mu_j}{dt}\right) - A_{ij}\rho_{,i}\left(\frac{d\mu_j}{dt}\right)$$

$$- A_{ij}\rho\left(\frac{d\mu_j}{dt}\right)_{,i} - f_j^M v_j - \rho\mu_j\frac{\partial B_j^M}{\partial t}, \qquad (5.3.17)$$

where Eq. (5.1.9) and the linear momentum equation in Eq. (5.3.5) have been used. With the use of Eq. (5.3.16), the energy equation can be written as

$$\rho\frac{d\varepsilon}{dt} = \tau_{ij}v_{j,i} - (B_j^M + B_j^L)\rho\frac{d\mu_j}{dt} - \rho A_{ij}\left(\frac{d\mu_j}{dt}\right)_{,i} - \rho\mu_j\frac{dB_j^M}{dt}. \qquad (5.3.18)$$

In summary, corresponding to Eqs. (5.2.1)–(5.2.7), we have

$$B_{k,k}^M = 0, \qquad (5.3.19)$$

$$\varepsilon_{ijk}H_{k,j} = 0, \qquad (5.3.20)$$

$$\rho^0 = \rho J, \qquad (5.3.21)$$

$$\tau_{ij,i} + \rho f_j + f_j^M = \rho\ddot{y}_j, \qquad (5.3.22)$$

$$\varepsilon_{ijk}\tau_{jk} - c_i^L = 0, \qquad (5.3.23)$$

$$\varepsilon_{ijk}\mu_j\left(-A_{lk,l} - \frac{A_{lk}}{\rho}\rho_{,l} + B_k^M + B_k^L\right) = \frac{1}{\gamma}\frac{d\mu_i}{dt}, \qquad (5.3.24)$$

$$\rho\frac{d\varepsilon}{dt} = \tau_{ij}v_{j,i} - (B_j^M + B_j^L)\rho\frac{d\mu_j}{dt} - \rho A_{ij}\left(\frac{d\mu_j}{dt}\right)_{,i} - \rho\mu_j\frac{dB_j^M}{dt}.$$

$$(5.3.25)$$

When applied to an interface between two materials or the boundary surface of a finite body, the integral balance laws lead to jump conditions or boundary conditions. On a boundary surface we may prescribe [17]:

$$\mathbf{y} \quad \text{or} \quad \mathbf{n} \cdot (\boldsymbol{\tau} + \mathbf{T}^M), \tag{5.3.26}$$

$$\psi \quad \text{or} \quad \mathbf{n} \cdot \mathbf{B}^M, \tag{5.3.27}$$

$$\delta\theta \quad \text{or} \quad \mathbf{n} \cdot \mathbf{A} \times \rho\boldsymbol{\mu}. \tag{5.3.28}$$

5.4 Constitutive Relations

We introduce an enthalpy density function χ per unit mass through the following Legendre transform:

$$\chi = \varepsilon + B_i^M \mu_i. \tag{5.4.1}$$

Then the energy equation in Eq. (5.3.25) becomes

$$\rho\frac{d\chi}{dt} = \tau_{ij}v_{j,i} - B_j^L \rho\frac{d\mu_j}{dt} - \rho A_{ij}\left(\frac{d\mu_j}{dt}\right)_{,i}, \tag{5.4.2}$$

or

$$\rho\frac{d\chi}{dt} = \tau_{ij}X_{M,i}\frac{d}{dt}(y_{j,M}) - \rho B_j^L \frac{d\mu_j}{dt} - \rho A_{ij}X_{M,i}\frac{d}{dt}(\mu_{j,M}), \tag{5.4.3}$$

where we have used

$$v_{j,i} = X_{M,i}\frac{d}{dt}(y_{j,M}), \quad \left(\frac{d\mu_j}{dt}\right)_{,i} = X_{M,i}\frac{d}{dt}(\mu_{j,M}). \tag{5.4.4}$$

According to Eq. (5.4.3), we let

$$\chi = \chi(y_{j,M}; \mu_i; \mu_{j,M}). \tag{5.4.5}$$

Hence,

$$\frac{d\chi}{dt} = \frac{\partial\chi}{\partial(y_{j,M})}\frac{d}{dt}(y_{j,M}) + \frac{\partial\chi}{\partial\mu_i}\frac{d\mu_i}{dt} + \frac{\partial\chi}{\partial(\mu_{j,M})}\frac{d}{dt}(\mu_{j,M}). \tag{5.4.6}$$

Substituting Eq. (5.4.6) into Eq. (5.4.3) and introducing the constraints in Eq. (5.1.7)$_{1,3}$ using Lagrange multipliers λ and L_M, we obtain

$$\rho\frac{\partial\chi}{\partial(y_{j,M})}\frac{d}{dt}(y_{j,M}) + \rho\frac{\partial\chi}{\partial\mu_i}\frac{d\mu_i}{dt} + \rho\frac{\partial\chi}{\partial(\mu_{j,M})}\frac{d}{dt}(\mu_{j,M})$$

$$= \tau_{ij}X_{M,i}\frac{d}{dt}(y_{j,M}) - \rho B_j^L \frac{d\mu_j}{dt} - \rho A_{ij}X_{M,i}\frac{d}{dt}(\mu_{j,M})$$

$$+ \lambda\rho\mu_i\frac{d\mu_i}{dt} + L_M\rho\left[\mu_{j,M}\frac{d\mu_j}{dt} + \mu_j\frac{d}{dt}(\mu_{j,M})\right], \tag{5.4.7}$$

or

$$\left[X_{M,i}\tau_{ij} - \rho\frac{\partial\chi}{\partial(y_{j,M})}\right]\frac{d}{dt}(y_{j,M}) - \rho\left[B_i^L - \lambda\mu_i - L_M\mu_{i,M} + \frac{\partial\chi}{\partial\mu_i}\right]\frac{d\mu_i}{dt}$$

$$- \rho\left[X_{M,i}A_{ij} - L_M\mu_j + \frac{\partial\chi}{\partial(\mu_{j,M})}\right]\frac{d}{dt}(\mu_{j,M}) = 0. \qquad (5.4.8)$$

Equation (5.4.8) implies the following constitutive relations:

$$X_{M,i}\tau_{ij} = \rho\frac{\partial\chi}{\partial(y_{j,M})},$$

$$B_i^L = -\frac{\partial\chi}{\partial\mu_i} + \lambda\mu_i + L_M\mu_{i,M}, \qquad (5.4.9)$$

$$X_{M,i}A_{ij} = -\frac{\partial\chi}{\partial(\mu_{j,M})} + L_M\mu_j.$$

Furthermore, $\mathbf{A}\cdot\mathbf{M} = 0$ implies that

$$X_{M,i}A_{ij}\mu_j = -\frac{\partial\chi}{\partial(\mu_{j,M})}\mu_j + L_M\mu_j\mu_j = 0, \qquad (5.4.10)$$

or

$$L_M = \frac{1}{\mu_s^2}\frac{\partial\chi}{\partial(\mu_{k,M})}\mu_k. \qquad (5.4.11)$$

In addition, $\mathbf{B}^L\cdot\mathbf{M} = 0$ implies that

$$B_i^L\mu_i = -\frac{\partial\chi}{\partial\mu_i}\mu_i + \lambda\mu_i\mu_i + L_M\mu_i\mu_{i,M} = 0, \qquad (5.4.12)$$

or

$$\lambda = \frac{1}{\mu_s^2}\frac{\partial\chi}{\partial\mu_k}\mu_k. \qquad (5.4.13)$$

With substitutions from Eqs. (5.4.11) and (5.4.13), the constitutive relations in Eq. (5.4.9) become

$$\tau_{ij} = \rho y_{i,M}\frac{\partial\chi}{\partial(y_{j,M})}, \qquad (5.4.14)$$

$$B_i^L = -\frac{\partial\chi}{\partial\mu_i} + \frac{1}{\mu_s^2}\left[\frac{\partial\chi}{\partial\mu_k}\mu_k\mu_i + \frac{\partial\chi}{\partial(\mu_{k,M})}\mu_k\mu_{i,M}\right], \qquad (5.4.15)$$

$$A_{ij} = -y_{i,M}\left[\frac{\partial\chi}{\partial(\mu_{j,M})} - \frac{1}{\mu_s^2}\frac{\partial\chi}{\partial(\mu_{k,M})}\mu_k\mu_j\right]. \qquad (5.4.16)$$

To satisfy rotational invariance (objectivity [1]) and Eq. (5.3.12), χ can be reduced to a function of the following inner products [16, 19]:

$$C_{KL} = y_{i,K} y_{i,L}, \quad G_{LM} = \mu_{i,K} \mu_{i,M}, \quad N_L = y_{i,L} \mu_i. \tag{5.4.17}$$

In addition, we will use E_{KL} instead of C_{KL}. Therefore, we take

$$\chi = \chi(E_{KL}; N_K; G_{LM}). \tag{5.4.18}$$

We assume that χ is written symmetrically in the following sense:

$$\frac{\partial \chi}{\partial E_{KL}} = \frac{\partial \chi}{\partial E_{LK}}, \quad \frac{\partial \chi}{\partial G_{KL}} = \frac{\partial \chi}{\partial G_{LK}}. \tag{5.4.19}$$

In differentiating χ, the elements of E_{KL} (or G_{KL}) are treated independently, i.e.,

$$\frac{\partial E_{KL}}{\partial E_{LK}} = 0, \quad \frac{\partial G_{KL}}{\partial G_{LK}} = 0, \quad K \neq L. \tag{5.4.20}$$

Then, it can be shown that

$$L_M = \frac{\partial \chi}{\partial(\mu_{k,M})} \mu_k = 0. \tag{5.4.21}$$

With Eq. (5.4.18), the constitutive relations in Eqs. (5.4.14)–(5.4.16) becomes

$$\tau_{ij} = \rho y_{i,M} \frac{\partial \chi}{\partial E_{ML}} y_{j,L} + \rho y_{i,M} \frac{\partial \chi}{\partial N_M} \mu_j, \tag{5.4.22}$$

$$B_i^L = -y_{i,L} \frac{\partial \chi}{\partial N_L} + \frac{1}{\mu_s^2} y_{k,L} \mu_k \mu_i \frac{\partial \chi}{\partial N_L}, \tag{5.4.23}$$

$$A_{ij} = -2 y_{i,M} \mu_{j,L} \frac{\partial \chi}{\partial G_{ML}}. \tag{5.4.24}$$

As an example, χ may be taken as [16]

$$\chi = \frac{1}{2\rho^0} c_{KLMN} E_{KL} E_{MN} + \frac{1}{2} \rho_2^0 \chi_{KL} N_K N_L$$

$$+ \frac{1}{3} (\rho^0)^2 {}_3\chi_{KLM} N_K N_L N_M + \frac{1}{4} (\rho^0)^3 {}_4\chi_{KLMN} N_K N_L N_M N_N$$

$$+ h_{KLM} N_K E_{LM} + \rho^0 b_{KLMN} N_K N_L E_{MN}$$

$$+ \rho^0 \gamma_{KLMN} G_{KL} E_{MN} + (\rho^0)^2 f_{KLM} N_K G_{LM} + \frac{1}{2} \rho^0 \alpha_{KL} G_{KL} \ldots, \tag{5.4.25}$$

where

c_{KLMN} — second order elastic constants,
$_2\chi_{KL}$ — second order anisotropy constants,
$_3\chi_{KLM}$ — third order anisotropy constants,
$_4\chi_{KLMN}$ — fourth order anisotropy constants,
h_{KLM} — piezomagnetic constants,
b_{KLMN} — magnetostrictive constants,
γ_{KLMN} — exchangestrictive constants,
f_{KLM} — magnetoexchange constants,
α_{KL} — exchange constants.

5.5 Small Deformation and Finite Magnetic Fields

The equations in the previous sections of this chapter are valid for finite deformations and finite magnetic fields. They are nonlinear both mechanically and magnetically. In the literature, mechanically linear theories for small deformations are often used [30, 31] while magnetically the equations are still nonlinear. In this section, we specialize the fully nonlinear equations in the previous sections to mechanically linear equations for small deformations. For small deformations, from Chapter 1, we have, approximately,

$$J \cong 1 + u_{m,m},$$
$$E_{KL} \to S_{kl} = \frac{1}{2}(u_{k,l} + u_{l,k}), \tag{5.5.1}$$

$$\rho = \frac{\rho^0}{J} \cong \frac{\rho^0}{1 + u_{m,m}} \cong \rho^0(1 - u_{m,m}),$$
$$M_k = \rho\mu_k \cong \rho^0(1 - u_{m,m})\mu_k, \tag{5.5.2}$$
$$N_L = y_{i,L}\mu_i = (\delta_{iL} + u_{i,L})\mu_i.$$

Our goal is to reduce the fully nonlinear formulation in the previous sections to what is similar to [30, 31], not a rigorous first-order theory of the small displacement gradients. Hence, we make some further approximations below which may be somewhat crude:

$$\rho \cong \rho^0(1 - u_{m,m}) \cong \rho^0,$$
$$M_k \cong \rho^0(1 - u_{m,m})\mu_k \cong \rho^0\mu_k, \tag{5.5.3}$$
$$N_L = (\delta_{iL} + u_{i,L})\mu_i \cong \delta_{iL}\mu_i.$$

Then the energy density in Eq. (5.4.25) may be approximated by

$$\rho^0 \chi \cong \frac{1}{2} c_{ijkl} S_{ij} S_{kl} + \frac{1}{2} 2\chi_{kl} M_k M_l$$

$$+ h_{klm} M_k S_{lm} + b_{klmn} M_k M_l S_{mn} + \frac{1}{2} \alpha_{kl} M_{m,k} M_{m,l}, \qquad (5.5.4)$$

where the terms associated with $_3\chi_{ijk}$, $_4\chi_{ijkl}$, γ_{klmn} and f_{lmn} are not considered. Equation (5.5.4) is similar to those in [30, 31]. From Eqs. (5.5.4) and (5.4.22)–(5.4.24), we obtain the corresponding constitutive relations as

$$\tau_{ij} = \rho^0 \frac{\partial \chi}{\partial S_{ij}} + \rho^0 \frac{\partial \chi}{\partial M_i} M_j = c_{ijkl} S_{kl} + h_{kij} M_k + b_{klij} M_k M_l$$

$$+ (_2\chi_{ik} M_k + h_{ilm} S_{lm} + 2b_{ikmn} M_k S_{mn}) M_j, \qquad (5.5.5)$$

$$B_i^L = -\rho^0 \frac{\partial \chi}{\partial M_i} + \frac{1}{M_s^2} M_k M_i \rho^0 \frac{\partial \chi}{\partial M_k}$$

$$= - (_2\chi_{ik} M_k + h_{ilm} S_{lm} + 2b_{ikmn} M_k S_{mn})$$

$$+ \frac{1}{M_s^2} M_k M_i (_2\chi_{kl} M_l + h_{klm} S_{lm} + 2b_{klmn} M_l S_{mn}), \qquad (5.5.6)$$

$$A_{ij} = -2\mu_{j,l} \frac{\partial \chi}{\partial G_{il}} = -2 \frac{1}{(\rho^0)^2} M_{j,l} \rho^0 \frac{\partial \chi}{\partial G_{il}}$$

$$= - 2 \frac{1}{(\rho^0)^2} M_{j,l} \rho^0 \frac{1}{2} \rho^0 \alpha_{il} = -\alpha_{ik} M_{j,k}. \qquad (5.5.7)$$

5.6 Thermal and Dissipative Effects

The equations in [16] include thermal and dissipative effects which were not considered in the previous sections of this chapter for simplicity. In this section, thermal and dissipative effects are treated for completeness. The balance laws in Eqs. (5.2.1)–(5.2.6) remain the same. The energy equation in Eq. (5.2.7) needs to be generalized to include thermal effects in the following manner:

$$\frac{d}{dt} \int_v \rho \left(\frac{1}{2} \mathbf{v} \cdot \mathbf{v} + \varepsilon \right) dv = \int_s \left(\mathbf{t} \cdot \mathbf{v} + \mathbf{F} \cdot \rho \frac{d\boldsymbol{\mu}}{dt} - \mathbf{n} \cdot \mathbf{q} \right) ds$$

$$+ \int_v \left[\rho \mathbf{f} \cdot \mathbf{v} - \rho \boldsymbol{\mu} \cdot \frac{\partial \mathbf{B}^M}{\partial t} + \rho r \right] dv, \qquad (5.6.1)$$

where r is the body heat source per unit mass and $\mathbf{q}$ is the heat flux vector. In addition, the second law of thermal dynamics needs to be added:

$$\frac{d}{dt}\int_v \rho\eta \, dv \geq \int_v \frac{\rho r}{\theta} \, dv - \int_s \frac{\mathbf{q}\cdot\mathbf{n}}{\theta} \, ds, \qquad (5.6.2)$$

where θ is the absolute temperature and η is the entropy density per unit mass.

Equations (5.6.1) and (5.6.2) can be brought into differential forms as

$$\rho\frac{d\varepsilon}{dt} = \tau_{ij}v_{j,i} - A_{ij,i}\rho\left(\frac{d\mu_j}{dt}\right) - A_{ij}\rho_{,i}\left(\frac{d\mu_j}{dt}\right)$$

$$- A_{ij}\rho\left(\frac{d\mu_j}{dt}\right)_{,i} - f_j^M v_j - \rho\mu_j\frac{\partial B_j^M}{\partial t} + \rho r - q_{i,i}, \qquad (5.6.3)$$

$$\rho\frac{d\eta}{dt} \geq \frac{\rho r}{\theta} - \left(\frac{q_i}{\theta}\right)_{,i}, \qquad (5.6.4)$$

where the linear momentum equation has been used. With the use of Eq. (5.3.16), the energy equation in Eq. (5.6.3) becomes

$$\rho\frac{d\varepsilon}{dt} = \tau_{ij}v_{j,i} - (B_j^M + B_j^L)\rho\frac{d\mu_j}{dt}$$

$$- \rho A_{ij}\left(\frac{d\mu_j}{dt}\right)_{,i} - \rho\mu_j\frac{dB_j^M}{dt} + \rho r - q_{i,i}. \qquad (5.6.5)$$

We introduce

$$\chi = \varepsilon + B_i^M\mu_i. \qquad (5.6.6)$$

Then the energy equation in Eq. (5.6.5) takes the following form:

$$\rho\frac{d\chi}{dt} = \tau_{ij}v_{j,i} - \rho A_{ij}\left(\frac{d\mu_j}{dt}\right)_{,i} - B_j^L\rho\frac{d\mu_j}{dt} + \rho r - q_{i,i}. \qquad (5.6.7)$$

The second law of thermodynamics in Eq. (5.6.4) can be further written as

$$\rho r \leq \theta\left(\frac{q_i}{\theta}\right)_{,i} + \theta\rho\frac{d\eta}{dt} = \theta\left(\frac{q_{i,i}}{\theta} - \frac{q_i}{\theta^2}\theta_{,i}\right) + \theta\rho\frac{d\eta}{dt} = q_{i,i} - \frac{q_i}{\theta}\theta_{,i} + \theta\rho\frac{d\eta}{dt}. \qquad (5.6.8)$$

Eliminating r from Eqs. (5.6.7) and (5.6.8), we obtain the Clausius–Duhem inequality as

$$\rho\frac{d\chi}{dt} - \tau_{ij}v_{j,i} + \rho A_{ij}\left(\frac{d\mu_j}{dt}\right)_{,i} + \rho B_j^L\frac{d\mu_j}{dt} + q_{i,i} \leq q_{i,i} - \frac{q_i}{\theta}\theta_{,i} + \theta\rho\frac{d\eta}{dt}, \qquad (5.6.9)$$

or

$$\rho\left(\theta\frac{d\eta}{dt} - \frac{d\chi}{dt}\right) + \tau_{ij}v_{j,i} - \rho A_{ij}\left(\frac{d\mu_j}{dt}\right)_{,i} - \rho B_j^L\frac{d\mu_j}{dt} - \frac{q_i}{\theta}\theta_{,i} \geq 0. \quad (5.6.10)$$

Under the following Legendre transform:

$$F = \chi - \theta\eta, \quad \chi = F + \theta\eta, \quad (5.6.11)$$

and hence

$$\frac{d\chi}{dt} = \frac{dF}{dt} + \frac{d\theta}{dt}\eta + \theta\frac{d\eta}{dt}, \quad (5.6.12)$$

the Clausius–Duhem inequality in Eq. (5.6.10) becomes

$$-\rho\left(\frac{dF}{dt} + \frac{d\theta}{dt}\eta\right) + \tau_{ij}v_{j,i} - \rho A_{ij}\left(\frac{d\mu_j}{dt}\right)_{,i} - \rho B_j^L\frac{d\mu_j}{dt} - \frac{q_i}{\theta}\theta_{,i} \geq 0.$$
$$(5.6.13)$$

At the same time the energy equation in Eq. (5.6.7) becomes

$$\rho\frac{dF}{dt} + \rho\frac{d\theta}{dt}\eta + \rho\theta\frac{d\eta}{dt} = \tau_{ij}v_{j,i} - \rho A_{ij}\left(\frac{d\mu_j}{dt}\right)_{,i} - \rho B_j^L\frac{d\mu_j}{dt} + \rho r - q_{i,i}.$$
$$(5.6.14)$$

For constitutive relations we break $\boldsymbol{\tau}$ and $\mathbf{B}^L$ into recoverable and dissipative parts through

$$\boldsymbol{\tau} = {}^R\boldsymbol{\tau} + {}^D\boldsymbol{\tau}, \quad \mathbf{B}^L = {}^R\mathbf{B}^L + {}^D\mathbf{B}^L. \quad (5.6.15)$$

The recoverable parts of Eq. (5.6.15) satisfy

$$\rho\frac{dF}{dt} = {}^R\tau_{ij}v_{j,i} - {}^RB_j^L\rho\frac{d\mu_j}{dt} - \rho A_{ij}\left(\frac{d\mu_j}{dt}\right)_{,i} - \rho\eta\frac{d\theta}{dt}. \quad (5.6.16)$$

Then the energy equation in Eq. (5.6.14) and the Clausius–Duhem inequality in Eq. (5.6.13) reduce to

$$\rho\theta\frac{d\eta}{dt} = {}^D\tau_{ij}v_{j,i} - {}^DB_j^L\rho\frac{d\mu_j}{dt} + \rho r - q_{i,i}, \quad (5.6.17)$$

$$ {}^D\tau_{ij}v_{j,i} - {}^DB_j^L\rho\frac{d\mu_j}{dt} - \frac{q_i}{\theta}\theta_{,i} \geq 0. \quad (5.6.18)$$

Equation (5.6.17) is the dissipation equation. With

$$v_{j,i} = X_{M,i}\frac{d}{dt}(y_{j,M}), \quad \left(\frac{d\mu_j}{dt}\right)_{,i} = X_{M,i}\frac{d}{dt}(\mu_{j,M}), \quad (5.6.19)$$

we write Eq. (5.6.16) as

$$\rho \frac{dF}{dt} = {}^{R}\tau_{ij} X_{M,i} \frac{d}{dt}(y_{j,M}) - {}^{R}B_{i}^{L} \rho \frac{d\mu_{i}}{dt} - \rho A_{ij} X_{M,i} \frac{d}{dt}(\mu_{j,M}) - \rho \eta \frac{d\theta}{dt}.$$

$$(5.6.20)$$

Based on Eq. (5.6.20), we let

$$F = F(y_{i,M}; \mu_{i}; \mu_{i,M}; \theta). \qquad (5.6.21)$$

When Eq. (5.6.21) is substituted into Eq. (5.6.20), in a way similar to the derivations in Section 5.4, it can be shown that the recoverable constitutive relations are given by

$$X_{M,i}{}^{R}\tau_{ij} = \rho \frac{\partial F}{\partial(y_{j,M})},$$

$$^{R}B_{i}^{L} = -\frac{\partial F}{\partial \mu_{i}} + \lambda \mu_{i} + L_{M}\mu_{i,M}, \qquad (5.6.22)$$

$$X_{M,i}A_{ij} = -\frac{\partial F}{\partial(\mu_{j,M})} + L_{M}\mu_{j},$$

$$\eta = -\frac{\partial F}{\partial \theta}. \qquad (5.6.23)$$

L_{M} and λ can be determined from Eqs. (5.1.4) and (5.3.11). The rotational invariance of F and Eq. (5.3.12) also need to be satisfied. These are similar to what are in Section 5.4. The constitutive relations for the dissipative parts of τ and $\mathbf{B}^{L}$ need to be given separately, along with a constitutive relation for $\mathbf{q}$. They also need to satisfy the rotational invariance and are restricted by the Clausius–Duhem inequality in Eq. (5.6.18).

Chapter 6

SMALL FERROMAGNETOELASTIC
FIELDS ON A FINITE BIAS

In many problems of ferromagnetic materials, the spontaneous magneti-zation may be treated as a biasing or initial field. If the material has magnetoelastic couplings, there may exist initial elastic deformations. Therefore, the theory of small deformations and fields superposed on a finite bias is used very often in ferromagnetoelastic materials. The linearization of the fully nonlinear theory of ferromagnetoelasticity in Chapter 5 about a finite bias is rather lengthy. The equations used in this chapter as well as Chapters 7 and 8 are taken directly from [16–18] in their notation with $\mathbf{H}$ instead of $\mathbf{B}$ in the magnetic body force and couple, and are in Gaussian units.

6.1 Small Deformation and Weak Fields on a Finite Bias

Consider the three states of the elastic ferromagnet shown in Fig. 6.1. The reference state is free from any deformation and fields. The initial state has an initial magnetization $\mathbf{M}^0$ which is finite and static. $\mathbf{M}^0$ is accompanied by initial magnetic fields and initial elastic deformations which are also static and finite. These initial deformations and fields are governed by the static and nonlinear equations of saturated ferromagnetoelastic solids [16], and are considered known. Then, small and dynamic loads are applied. The magnetization in the present state is $\mathbf{M}$. The deformations and magnetic fields at the present state satisfy the nonlinear and dynamic equations of saturated ferromagnetoelastic solids [16]. The small and incremental deformations and fields between the initial and present states are as follows: $\mathbf{u}$ is the displacement. $\boldsymbol{\tau}$ is the stress tensor which has a symmetric part $\boldsymbol{\tau}^S$ and an antisymmetric part $\boldsymbol{\tau}^A$. $\mathbf{f}$ is the mechanical body force. ψ is the magnetostatic potential. $\mathbf{h}^M$ is the Maxwellian magnetic field. $\mathbf{h}^L$ is the effective local magnetic field which describes the interaction between

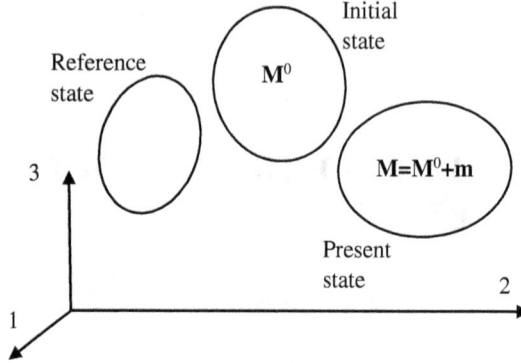

Fig. 6.1.   Reference, initial and present states.

the spin and the lattice. $\mathbf{b}^M$ is the magnetic induction. $\mathbf{a}$ is the exchange interaction between neighboring spins. $\mathbf{m}$ is the magnetization.

The linear equations governing the incremental, small and dynamic fields on top of the static and finite bias are obtained by linearizing the nonlinear and dynamic equations about the bias [16, 18, 32, 33]:

$$\tau_{ij,i} + f_j + M_i^0 h_{j,i}^M = \rho \ddot{u}_j, \tag{6.1.1}$$

$$\tau_{ij}^A = \frac{1}{2}(M_i^0 h_j^L - M_j^0 h_i^L), \tag{6.1.2}$$

$$\varepsilon_{ijk} h_{k,j}^M = 0, \tag{6.1.3}$$

$$b_{i,i}^M = 0, \tag{6.1.4}$$

$$\varepsilon_{ijk} M_j^0 (h_k^M - a_{lk,l} + h_k^L) + \varepsilon_{ijk} m_j H_k^0 = \frac{1}{\gamma} \dot{m}_i. \tag{6.1.5}$$

Equation (6.1.1) is the linear momentum equation of the combined continuum. Equation (6.1.2) is the angular momentum equation of the lattice continuum. Equations (6.1.3) and (6.1.4) are the relevant ones from Maxwell's equations for the quasistatic magnetic field and induction. Equation (6.1.5) is the angular momentum equation of the spin continuum. γ is the gyromagnetic ratio which is a negative number. The constitutive relations are [16, 18]

$$
\begin{aligned}
\tau_{ij}^S &= \bar{c}_{ijkm} u_{k,m} + \bar{h}_{kij} m_k, \\
h_i^L &= \bar{\chi}_{ik} m_k + \bar{g}_{ikm} u_{k,m}, \\
a_{ij} &= \bar{\beta}_{ijkl} m_{l,k},
\end{aligned}
\tag{6.1.6}
$$

where $\bar{c}_{ijkm}$ are the effective elastic constants, $\bar{\chi}_{ik}$ the effective magnetic anisotropy constants, $\bar{h}_{kij}$ and $\bar{g}_{ikm}$ the effective piezomagnetic constants, and $\bar{\beta}_{ijkl}$ the effective exchange constants. Their general expressions in terms of the fundamental material constants and initial fields can be found in [16, 18]. We also have the following relationships [16, 18]:

$$\tau_{ij} = \tau_{ij}^S + \tau_{ij}^A. \tag{6.1.7}$$

$$b_i^M = h_i^M + 4\pi m_i - 4\pi M_i^0 u_{j,j}. \tag{6.1.8}$$

Equation (6.1.3) is satisfied with the magnetostatic potential ψ through

$$h_i^M = -\psi_{,i}. \tag{6.1.9}$$

Then, in component form, Eqs. (6.1.1), (6.1.4) and (6.1.5) can be written as seven equations for ψ and the components of **u** and **m**.

6.2　Equations for Cubic Crystals

Consider cubic crystals of class (m3m). These crystals are nonpiezomagnetic in their reference state free from any deformations and fields, i.e., $h_{klm} = 0$. The other relevant material matrices are [18, 20]

$$[c_{pq}] = \begin{bmatrix} c_{11} & c_{12} & c_{12} & 0 & 0 & 0 \\ c_{12} & c_{11} & c_{12} & 0 & 0 & 0 \\ c_{12} & c_{12} & c_{11} & 0 & 0 & 0 \\ 0 & 0 & 0 & c_{44} & 0 & 0 \\ 0 & 0 & 0 & 0 & c_{44} & 0 \\ 0 & 0 & 0 & 0 & 0 & c_{44} \end{bmatrix}, \tag{6.2.1}$$

$$[b_{pq}] = \begin{bmatrix} b_{11} & b_{12} & b_{12} & 0 & 0 & 0 \\ b_{12} & b_{11} & b_{12} & 0 & 0 & 0 \\ b_{12} & b_{12} & b_{11} & 0 & 0 & 0 \\ 0 & 0 & 0 & b_{44} & 0 & 0 \\ 0 & 0 & 0 & 0 & b_{44} & 0 \\ 0 & 0 & 0 & 0 & 0 & b_{44} \end{bmatrix}, \tag{6.2.2}$$

$$[4\chi_{pq}] = \begin{bmatrix} 4\chi_{11} & 4\chi_{12} & 4\chi_{12} & 0 & 0 & 0 \\ 4\chi_{12} & 4\chi_{11} & 4\chi_{12} & 0 & 0 & 0 \\ 4\chi_{12} & 4\chi_{12} & 4\chi_{11} & 0 & 0 & 0 \\ 0 & 0 & 0 & 4\chi_{12} & 0 & 0 \\ 0 & 0 & 0 & 0 & 4\chi_{12} & 0 \\ 0 & 0 & 0 & 0 & 0 & 4\chi_{12} \end{bmatrix}, \tag{6.2.3}$$

$$[_2\chi_{ij}] = \begin{bmatrix} 2\chi_{11} & 0 & 0 \\ 0 & 2\chi_{11} & 0 \\ 0 & 0 & 2\chi_{11} \end{bmatrix}, \tag{6.2.4}$$

$$[\alpha_{ij}] = \begin{bmatrix} \alpha_{11} & 0 & 0 \\ 0 & \alpha_{11} & 0 \\ 0 & 0 & \alpha_{11} \end{bmatrix}. \tag{6.2.5}$$

Specifically, for yttrium iron garnet or YIG ($Y_3Fe_5O_{12}$), the relevant material constants are [18]:

$$\rho = 5.172\,\text{g/cm}^3, \quad c_{11} = 26.9 \times 10^{11}\,\text{dyn/cm}^2,$$
$$c_{12} = 10.77 \times 10^{11}\,\text{dyn/cm}^2, \quad c_{44} = 7.64 \times 10^{11}\,\text{dyn/cm}^2, \tag{6.2.6}$$

$$b_{11} - b_{12} = 1.66 \times 10^2, \quad b_{44} = 1.66 \times 10^2,$$
$$\chi = _3{}_4\chi_{12} - _4\chi_{11} = 3.36 \times 10^{-5}\,\text{Oe}^{-2}, \tag{6.2.7}$$

$$\alpha_{11} = 1.87 \times 10^{-11}\,\text{cm}^2,$$
$$\gamma = -1.76 \times 10^7\,\text{Oe-cm}^2/\text{dyn-sec}, \tag{6.2.8}$$
$$M^0 = 1750/4\pi\,\text{G}.$$

In Eq. (6.2.7) and the rest of this chapter as well as Chapters 7 and 8, χ is a combination of material constants. It depends on the fourth-order magnetic anisotropy constants $_4\chi_{pq}$. Let the initial magnetization $\mathbf{M}^0$ be along the x_3 axis and $\mathbf{M} = \mathbf{M}^0 + \mathbf{m}$ where $\mathbf{m}$ is small. In this case $m_3 = 0$ because of the saturation condition $\mathbf{M} \cdot \mathbf{M} = 0$ which implies that $\mathbf{M}^0 \cdot \mathbf{m} = 0$. In addition, $\boldsymbol{\tau}^A$ is relatively small and is found to be negligible compared to $\boldsymbol{\tau}^S$ [18]. Then the effective constitutive relations for the incremental fields are [18]

$$\tau_1 = \tau_{11} = c_{11}u_{1,1} + c_{12}u_{2,2} + c_{12}u_{3,3},$$
$$\tau_2 = \tau_{22} = c_{12}u_{1,1} + c_{11}u_{2,2} + c_{12}u_{3,3}, \tag{6.2.9}$$
$$\tau_3 = \tau_{33} = c_{12}u_{1,1} + c_{12}u_{2,2} + c_{11}u_{3,3},$$

$$\tau_4 = \tau_{23}^S = c_{44}(u_{2,3} + u_{3,2}) + 2b_{44}M^0 m_2,$$
$$\tau_5 = \tau_{31}^S = c_{44}(u_{1,3} + u_{3,1}) + 2b_{44}M^0 m_1, \tag{6.2.10}$$
$$\tau_6 = \tau_{12}^S = c_{44}(u_{1,2} + u_{2,1}),$$

$$h_1^L = -\chi(M^0)^2 m_1 - 2b_{44}M^0(u_{1,3} + u_{3,1}),$$
$$h_2^L = -\chi(M^0)^2 m_2 - 2b_{44}M^0(u_{2,3} + u_{3,2}), \qquad (6.2.11)$$
$$h_3^L = 0,$$

$$a_{ib} = -2\alpha_{11}m_{b,i}. \qquad (6.2.12)$$

Equations $(6.2.10)_{1,2}$ show that under $\mathbf{M}^0$ the crystal becomes effectively piezomagnetic through the magnetostrictive constant b_{44}. The substitution of Eqs. $(6.2.9)$–$(6.2.12)$ into Eqs. $(6.1.1)$, $(6.1.5)$ and $(6.1.4)$ yields six equations for $\mathbf{u}$, m_1, m_2 and ψ as follows:

$$c_{11}u_{1,11} + c_{12}u_{2,21} + c_{12}u_{3,31}$$
$$+ c_{44}u_{1,22} + c_{44}u_{2,12} + c_{44}u_{1,33} + c_{44}u_{3,13} \qquad (6.2.13)$$
$$+ 2b_{44}M^0 m_{1,3} - M^0\psi_{,13} + f_1 = \rho\ddot{u}_1,$$

$$c_{11}u_{2,22} + c_{12}u_{1,12} + c_{12}u_{3,32}$$
$$+ c_{44}u_{2,11} + c_{44}u_{1,21} + c_{44}u_{2,33} + c_{44}u_{3,23} \qquad (6.2.14)$$
$$+ 2b_{44}M^0 m_{2,3} - M^0\psi_{,23} + f_2 = \rho\ddot{u}_2,$$

$$c_{44}u_{1,31} + c_{44}u_{3,11} + c_{44}u_{2,32} + c_{44}u_{3,22}$$
$$+ c_{12}u_{1,13} + c_{12}u_{2,23} + c_{11}u_{3,33} \qquad (6.2.15)$$
$$+ 2b_{44}M^0 m_{1,1} + 2b_{44}M^0 m_{2,2} - M^0\psi_{,33} + f_3 = \rho\ddot{u}_3,$$

$$M^0\psi_{,2} - 2\alpha_{11}M^0 m_{2,kk} + \chi(M^0)^3 m_2$$
$$+ 2b_{44}(M^0)^2(u_{2,3} + u_{3,2}) + H^0 m_2 = \frac{1}{\gamma}\dot{m}_1, \qquad (6.2.16)$$

$$-M^0\psi_{,1} + 2\alpha_{11}M^0 m_{1,kk} - \chi(M^0)^3 m_1$$
$$- 2b_{44}(M^0)^2(u_{1,3} + u_{3,1}) - H^0 m_1 = \frac{1}{\gamma}\dot{m}_2, \qquad (6.2.17)$$

$$\psi_{,kk} - 4\pi(m_{1,1} + m_{2,2}) + 4\pi M^0 u_{k,k3} = 0. \qquad (6.2.18)$$

6.3 Plane Waves

Consider plane waves propagating in the x_3 direction without x_1 and x_2 dependence. Equations $(6.2.13)$–$(6.2.18)$ reduce to two uncoupled groups. One is for effective piezomagnetic waves which are longitudinal:

$$c_{11}u_{3,33} - M^0\psi_{,33} = \rho\ddot{u}_3, \qquad (6.3.1)$$

$$\psi_{,33} + 4\pi M^0 u_{3,33} = 0. \qquad (6.3.2)$$

The other is for coupled spin waves and transverse elastic waves:

$$c_{44}u_{1,33} + 2b_{44}M^0 m_{1,3} = \rho\ddot{u}_1, \tag{6.3.3}$$

$$c_{44}u_{2,33} + 2b_{44}M^0 m_{2,3} = \rho\ddot{u}_2, \tag{6.3.4}$$

$$-2a_{11}M^0 m_{2,33} + [\chi(M^0)^3 + H^0]m_2 + 2b_{44}(M^0)^2 u_{2,3} = \frac{1}{\gamma}\dot{m}_1, \tag{6.3.5}$$

$$2a_{11}M^0 m_{1,33} - [\chi(M^0)^3 + H^0]m_1 - 2b_{44}(M^0)^2 u_{1,3} = \frac{1}{\gamma}\dot{m}_2. \tag{6.3.6}$$

For the piezomagnetic waves governed by Eqs. (6.3.1) and (6.3.2), let

$$\begin{bmatrix} u_3 \\ \psi \end{bmatrix} = \begin{bmatrix} U \\ \Psi \end{bmatrix} \exp[i(\zeta x_3 - \omega t)]. \tag{6.3.7}$$

The substitution of Eq. (6.3.7) into Eqs. (6.3.1) and (6.3.2) results in

$$\begin{aligned} c_{11}\zeta^2 U - M^0\zeta^2\Psi &= \rho\omega^2 U, \\ \zeta^2\Psi + 4\pi M^0\zeta^2 U &= 0. \end{aligned} \tag{6.3.8}$$

From Eq. $(6.3.8)_2$,

$$\Psi = -4\pi M^0 U. \tag{6.3.9}$$

Substituting Eq. (6.3.9) into Eq. $(6.3.8)_1$, we obtain the wave speed as

$$v = \frac{\omega}{\zeta} = \sqrt{\frac{c'_{11}}{\rho}}, \tag{6.3.10}$$

where c'_{11} is a piezomagnetically-stiffened elastic constant:

$$c'_{11} = c_{11} + 4\pi(M^0)^2. \tag{6.3.11}$$

For the coupled elastic and spin waves governed by Eqs. (6.3.3)–(6.3.6), let [18]

$$\begin{aligned} u_1 &= U_1 \sin\zeta x_3 \cos\omega t, \quad u_2 = U_2 \sin\zeta x_3 \sin\omega t, \\ m_1 &= C_1 \cos\zeta x_3 \cos\omega t, \quad m_2 = C_2 \cos\zeta x_3 \sin\omega t, \end{aligned} \tag{6.3.12}$$

which satisfies Eqs. (6.3.3)–(6.3.6) provided that

$$\begin{aligned} (c_{44}\zeta^2 - \rho\omega^2)U_1 + e\zeta C_1 &= 0, \\ (c_{44}\zeta^2 - \rho\omega^2)U_2 + e\zeta C_2 &= 0, \\ e\zeta U_2 + (\omega/M^0\gamma)C_1 + (\alpha\zeta^2 + P)C_2 &= 0, \\ e\zeta U_1 + (\alpha\zeta^2 + P)C_1 + (\omega/M^0\gamma)C_2 &= 0, \end{aligned} \tag{6.3.13}$$

where we have denoted

$$e = 2b_{44}M^0, \quad \alpha = 2\alpha_{11},$$
$$P = H^0/M^0 + K, \quad K = \chi(M^0)^2. \tag{6.3.14}$$

For nontrivial solutions,

$$\begin{vmatrix} c_{44}\zeta^2 - \rho\omega^2 & 0 & e\zeta & 0 \\ 0 & c_{44}\zeta^2 - \rho\omega^2 & 0 & e\zeta \\ 0 & e\zeta & \omega/M^0\gamma & \alpha\zeta^2 + P \\ e\zeta & 0 & \alpha\zeta^2 + P & \omega/M^0\gamma \end{vmatrix} = 0, \tag{6.3.15}$$

from which we obtain

$$\alpha c_{44}\zeta^4 + \left(Pc_{44} - e^2 - \alpha\rho\omega^2 \pm \frac{c_{44}}{\gamma M^0}\omega\right)\zeta^2$$
$$-P\rho\omega^2 \mp \frac{\rho}{\gamma M^0}\omega^3 = 0. \tag{6.3.16}$$

The dispersion relations determined from Eq. (6.3.16) are shown in Fig. 6.2.

In the special case of $b_{44} = 0$, the coupling between elastic and spin waves in Eqs. (6.3.3)–(6.3.6) disappears. In this case, we have uncoupled elastic waves with

$$\omega = \sqrt{\frac{c_{44}}{\rho}}\zeta, \tag{6.3.17}$$

and uncoupled spin waves with

$$\frac{\omega}{\gamma M^0} = \pm(\alpha\zeta^2 + P). \tag{6.3.18}$$

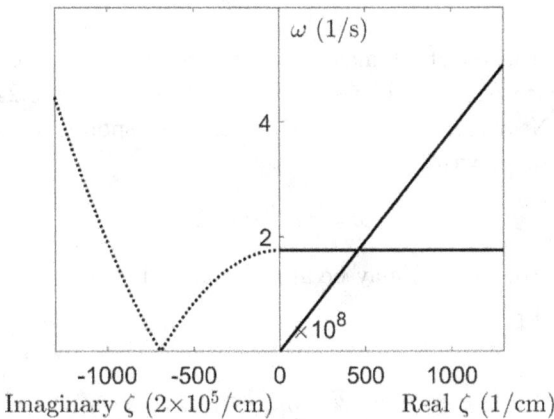

Fig. 6.2. Dispersion curves.

When $\zeta = 0$, the spin waves have the following cutoff frequency:

$$\omega|_{\zeta=0} = \pm\gamma M^0 P. \tag{6.3.19}$$

For long waves with a small ζ, Eq. (6.3.16) for coupled elastic and spin waves may be approximated by

$$\left(Pc_{44} - e^2 - \alpha\rho\omega^2 \pm \frac{c_{44}}{\gamma M^0}\omega\right)\zeta^2 - P\rho\omega^2 \mp \frac{\rho}{\gamma M^0}\omega^3 = 0. \tag{6.3.20}$$

For cutoff frequencies we set $\zeta = 0$ in Eq. (6.3.20) and it reduces to

$$-P\rho\omega^2 \mp \frac{\rho}{\gamma M^0}\omega^3 = 0. \tag{6.3.21}$$

The two roots of Eq. (6.3.21) are

$$\begin{aligned} \omega &= 0, \\ \omega &= \pm\gamma M^0 P, \end{aligned} \tag{6.3.22}$$

which are the same as the predictions from Eqs. (6.3.17) and (6.3.18) for uncoupled waves. Near $(\zeta, \omega) = (0, 0)$, Eq. (6.3.20) may be approximated by

$$(Pc_{44} - e^2)\zeta^2 - P\rho\omega^2 = 0, \tag{6.3.23}$$

or

$$\frac{\omega}{\zeta} = \sqrt{\frac{Pc_{44} - e^2}{P\rho}}, \tag{6.3.24}$$

which shows the effect of magnetoelastic coupling on elastic waves. In the special case of $b_{44} = 0$, we have $e = 0$ and Eq. (6.3.24) reduces to Eq. (6.3.17). Near $(\zeta, \omega) = (0, \gamma M^0 P)$ which corresponds to the lower signs in Eq. (6.3.20), we write

$$\omega = \gamma M^0 P + \Delta. \tag{6.3.25}$$

For small Δ, Eq. (6.3.20) may be approximated by

$$\begin{aligned} &\left[Pc_{44} - e^2 - \alpha\rho(\gamma M^0 P)^2 - \frac{c_{44}}{\gamma M^0}(\gamma M^0 P)\right]\zeta^2 \\ &- P\rho[(\gamma M^0 P)^2 + 2(\gamma M^0 P)\Delta] \\ &+ \frac{\rho}{\gamma M^0}\left[(\gamma M^0 P)^3 + 3(\gamma M^0 P)^2\Delta\right] = 0. \end{aligned} \tag{6.3.26}$$

Equation (6.3.26) leads to

$$\Delta = \frac{e^2 + \alpha\rho(\gamma M^0 P)^2}{P\rho(\gamma M^0 P)}\zeta^2. \tag{6.3.27}$$

Hence,

$$\omega = \gamma M^0 P + \frac{e^2 + \alpha\rho(\gamma M^0 P)^2}{P\rho(\gamma M^0 P)}\zeta^2. \tag{6.3.28}$$

Equation (6.3.28) describes long spin waves and shows the effect of magnetoelastic coupling. In the special case of $b_{44} = 0$ and hence $e = 0$, Eq. (6.3.28) reduces to

$$\omega = \gamma M^0(P + \alpha\zeta^2), \tag{6.3.29}$$

which agrees with Eq. (6.3.18).

6.4 Waves in Plates

In this section, we analyze waves propagating in the unbounded YIG plate in Fig. 6.3. The plate is free from mechanical loads. Its surfaces are the so-called perfect magnetic walls where the magnetic potential ψ vanishes. The plate surfaces are also assumed to be with $m_{1,3} = m_{2,3} = 0$. We consider straight-crested waves propagating in the x_1 direction without x_2 dependence. All fields have a common factor of $\exp(i\omega t - \xi x_1)$. Equations (6.2.13)–(6.2.18) reduce to

$$\begin{aligned} c_{11}u_{1,11} + c_{12}u_{3,13} &+ c_{44}(u_{1,33} + u_{3,13}) \\ &+ 2b_{44}M^0 m_{1,3} - M^0\psi_{,13} = \rho\ddot{u}_1, \end{aligned} \tag{6.4.1}$$

$$c_{44}u_{2,11} + c_{44}u_{2,33} + 2b_{44}M^0 m_{2,3} = \rho\ddot{u}_2, \tag{6.4.2}$$

$$\begin{aligned} c_{44}(u_{1,13} + u_{3,11}) &+ c_{12}u_{1,13} + c_{11}u_{3,33} \\ &+ 2b_{44}M^0 m_{1,1} - M^0\psi_{,33} = \rho\ddot{u}_3, \end{aligned} \tag{6.4.3}$$

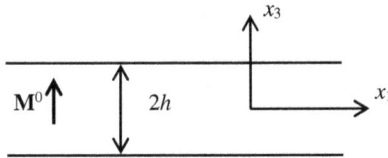

Fig. 6.3. A YIG plate and coordinate system.

$$M^0[-2\alpha_{11}m_{2,11} - 2\alpha_{11}m_{2,33} + \chi(M^0)^2 m_2$$
$$+ 2b_{44}M^0(u_{1,3} + u_{3,1})] + m_2 H^0 = \frac{1}{\gamma}\dot{m}_1, \tag{6.4.4}$$

$$-M^0[\psi_{,1} - 2\alpha_{11}m_{1,11} - 2\alpha_{11}m_{1,33} + \chi(M^0)^2 m_1$$
$$+ 2b_{44}M^0 u_{2,3}] - m_1 H^0 = \frac{1}{\gamma}\dot{m}_2, \tag{6.4.5}$$

$$-\psi_{,11} + 4\pi m_{1,1} - \psi_{,33} - 4\pi M^0(u_{1,13} + u_{3,33}) = 0. \tag{6.4.6}$$

Since the equation for the magnetic potential in Eq. (6.4.6) does not describe propagating waves and that ψ vanishes at the plate surfaces, we make an approximation by setting $\psi = 0$ and dropping Eq. (6.4.6). Then the so-called semi-analytical finite element method is used to obtain the dispersion curves in Figs. 6.4 and 6.5, where we have introduced the following dimensionless frequency and dimensionless wave number:

$$\Omega = \omega \bigg/ \left(\frac{\pi}{2c}\sqrt{\frac{c_{44}}{\rho}}\right), \quad X = \xi / \left(\frac{\pi}{2c}\right). \tag{6.4.7}$$

The dispersion curves of these waves separate into two groups. In one, u_1 and u_2 are odd functions of x_3 but u_3, m_1 and m_2 are even functions,

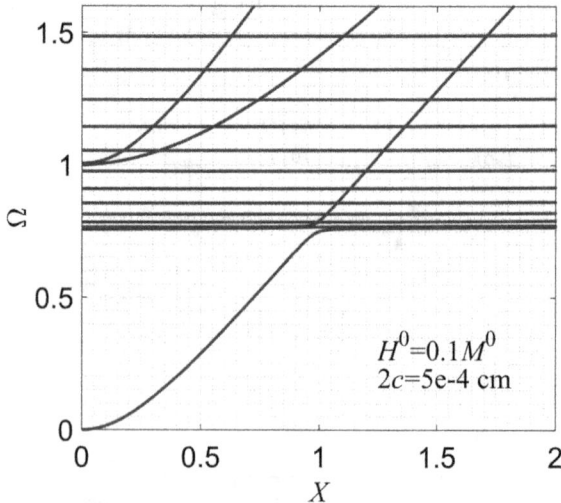

Fig. 6.4. Dispersion curves of plate waves. u_1 and u_2 are odd functions of x_3. u_3, m_1 and m_2 are even functions. $2h = 0.1\,\text{cm}$.

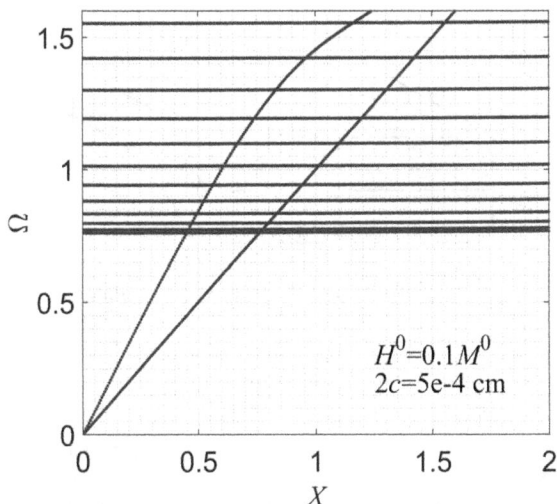

Fig. 6.5. Dispersion curves of plate waves. u_1 and u_2 are even functions of x_3. u_3, m_1 and m_2 are odd functions. $2h = 0.1\,\mathrm{cm}$.

which is shown in Fig. 6.4. The other group has the opposite symmetry and is shown in Fig. 6.5. In both Figs. 6.4 and 6.5, the nearly horizontal lines are the dispersion curves of spin waves (see Fig. 4.6). In Fig. 6.4, there are three branches of elastic waves. The one that goes through the origin represents flexural waves (u_3) which shows interaction with the lowest branch of spin waves. The other two branches are the first-order thickness-shear (u_1) and thickness-twist (u_2) waves. In Fig. 6.5, there are two branches for elastic waves. They both go through the origin and represent face-shear (u_2) and extensional waves (u_1).

6.5 Static Bending of a Rectangular Plate

In this section, we study the static bending of a rectangular YIG plate as shown in Fig. 6.6 under a mechanical load [34]. The plate is within $0 < x_1 < a$ and $0 < x_2 < b$. The thickness is given by $-h < x_3 < h$. The governing equations are Eqs. (6.2.13)–(6.2.18). The body force $\mathbf{f} = 0$. The two edges at $x_1 = 0$ or a and the two edges at $x_2 = 0$ or b are perfect magnetic walls with $\psi = 0$. We consider the following edge conditions which may be considered as a simply-supported plate with $u_3 = 0$

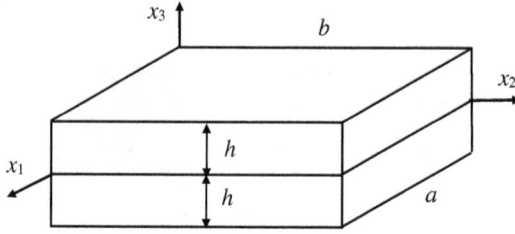

Fig. 6.6. A rectangular YIG plate and coordinate system.

and $\tau_{11} = 0$ or $\tau_{22} = 0$:

$$\tau_{11} = 0, \quad u_2 = 0, \quad u_3 = 0, \quad x_1 = 0, a,$$
$$\psi = 0, \quad m_{1,1} = 0, \quad m_2 = 0, \quad x_1 = 0, a, \tag{6.5.1}$$

$$u_1 = 0, \quad \tau_{22} = 0, \quad u_3 = 0, \quad x_2 = 0, b,$$
$$\psi = 0, \quad m_1 = 0, \quad m_{2,2} = 0, \quad x_2 = 0, b. \tag{6.5.2}$$

We look for a trigonometric series solution in the following form:

$$\{u_1, m_1\} = \sum_{m,n}^{\infty} \{U_{mn}(x_3), P_{mn}(x_3)\} \cos(\xi_m x_1) \sin(\eta_n x_2),$$

$$\{u_2, m_2\} = \sum_{m,n}^{\infty} \{V_{mn}(x_3), Q_{mn}(x_3)\} \sin(\xi_m x_1) \cos(\eta_n x_2), \tag{6.5.3}$$

$$\{u_3, \psi\} = \sum_{m,n}^{\infty} \{W_{mn}(x_3), \Psi_{mn}(x_3)\} \sin(\xi_m x_1) \sin(\eta_n x_2),$$

where

$$\xi_m = \frac{m\pi}{a}, \quad \eta_n = \frac{n\pi}{b}, \quad m, n = 1, 2, 3 \dots. \tag{6.5.4}$$

Equation (6.5.3) satisfies Eqs. (6.5.1) and (6.5.2). Substituting Eq. (6.5.3) into Eqs. (6.2.13)–(6.2.18), we obtain six ordinary differential equations for $U_{mn}, V_{mn}, W_{mn}, P_{mn}, Q_{mn}$ and Ψ_{mn}:

$$c_{44}U''_{mn} - (c_{11}\xi_m^2 + c_{44}\eta_n^2)U_{mn} - (c_{12} + c_{44})\xi_m\eta_n V_{mn}$$
$$+ (c_{12} + c_{44})\xi_m W'_{mn} + 2b_{44}M^0 P'_{mn} - M^0 \xi_m \Psi'_{mn} = 0, \tag{6.5.5}$$

$$c_{44}V''_{mn} - (c_{11}\eta_n^2 + c_{44}\xi_m^2)V_{mn} - (c_{12} + c_{44})\xi_m\eta_n U_{mn}$$
$$+ (c_{12} + c_{44})\eta_n W'_{mn} + 2b_{44}M^0 Q'_{mn} - M^0 \eta_n \Psi'_{mn} = 0, \tag{6.5.6}$$

$$c_{11}W''_{mn} - c_{44}(\xi_m^2 + \eta_n^2)W_{mn} - (c_{12} + c_{44})(\xi_m U'_{mn} + \eta_n V'_{mn})$$
$$-2b_{44}M^0(P_{mn}\xi_m + Q_{mn}\eta_n) - M^0\Psi''_{mn} = 0, \qquad (6.5.7)$$

$$-2M^0 P''_{mn}\alpha_{11} + [(M^0)^3\chi + 2M^0\alpha_{11}(\xi_m^2 + \eta_n^2) + H^0]P_{mn}$$
$$+2b_{44}\xi_m(M^0)^2 W_{mn} + 2b_{44}(M^0)^2 U'_{mn} + M^0\xi_m\Psi_{mn} = 0, \qquad (6.5.8)$$

$$-2M^0 Q''_{mn}\alpha_{11} + [(M^0)^3\chi + 2M^0\alpha_{11}(\xi_m^2 + \eta_n^2) + H^0]Q_{mn}$$
$$+2b_{44}\eta_n(M^0)^2 W_{mn} + 2b_{44}(M^0)^2 V'_{mn} + M^0\eta_n\Psi_{mn} = 0, \qquad (6.5.9)$$

$$\Psi''_{mn} - (\xi_m^2 + \eta_n^2)\Psi_{mn} - 4\pi M^0(\xi_m U'_{mn} + \eta_n V'_{mn} + W''_{mn})$$
$$+4\pi(\xi_m P_{mn} + \eta_n Q_{mn}) = 0. \qquad (6.5.10)$$

They are solved with the following boundary conditions at $x_3 = \pm h$:

$$\tau_{31}(\pm h) = \sum_{m,n}^{\infty} A_{mn}^{(1)\pm}\cos(\xi_m x_1)\sin(\eta_n x_2),$$

$$\tau_{32}(\pm h) = \sum_{m,n}^{\infty} A_{mn}^{(2)\pm}\sin(\xi_m x_1)\cos(\eta_n x_2), \qquad (6.5.11)$$

$$\tau_{33}(\pm h) = \sum_{m,n}^{\infty} A_{mn}^{(3)\pm}\sin(\xi_m x_1)\sin(\eta_n x_2),$$

$$b_3^M(\pm h) = \sum_{m,n}^{\infty} B_{mn}^{\pm}\sin(\xi_m x_1)\sin(\eta_n x_2), \qquad (6.5.12)$$

$$m_{1,3}(\pm h) = \sum_{m,n}^{\infty} C_{mn}^{(1)\pm}\cos(\xi_m x_1)\sin(\eta_n x_2),$$

$$m_{2,3}(\pm h) = \sum_{m,n}^{\infty} C_{mn}^{(2)\pm}\sin(\xi_m x_1)\cos(\eta_n x_2), \qquad (6.5.13)$$

where $A_{mn}^{(1)\pm}$, $A_{mn}^{(2)\pm}$, $A_{mn}^{(3)\pm}$, $B_{mn}^{\pm}$, $C_{mn}^{(1)\pm}$ and $C_{mn}^{(2)\pm}$ are known from the applied load at the plate top and bottom surfaces.

As an example, we consider the case when the top surface at $x_3 = h$ has a uniform normal load $\tau_{33}(h) = p$ within a small rectangular area centered at (x_0, y_0) as shown in Fig. 6.7. The rest of the top and bottom surfaces are without mechanical and magnetic loads. The only surface load applied can be represented by

$$\tau_{33}(h) = \begin{cases} p, & |x_1 - x_0| < c, \quad |x_2 - y_0| < d, \\ 0, & \text{elsewhere.} \end{cases} \qquad (6.5.14)$$

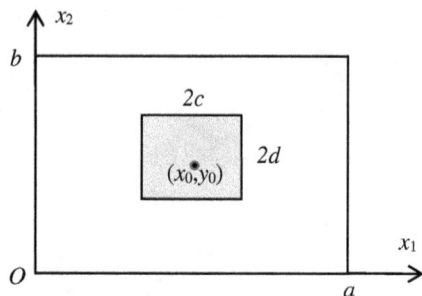

Fig. 6.7. Loading area centered at (x_0, y_0).

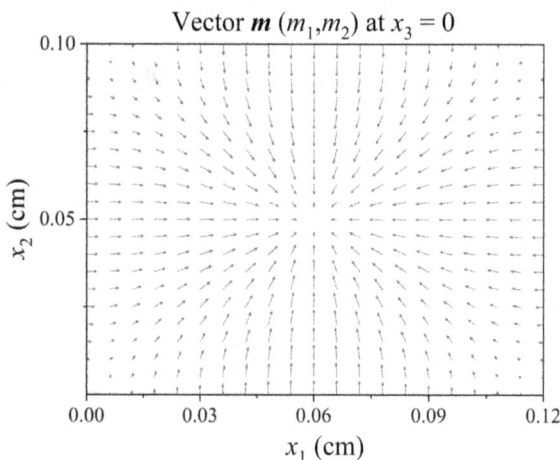

Fig. 6.8. Distribution of **m** at $x_3 = 0$.

Corresponding to Eq. (6.5.14), the coefficients in Eqs. (6.5.11)–(6.5.13) are

$$A_{mn}^{(3)+} = \frac{4}{ab}\frac{p}{\xi_m \eta_n}[\cos\xi_m(x_0 - c) - \cos\xi_m(x_0 + c)]$$
$$\times [\cos\eta_n(y_0 - d) - \cos\eta_n(y_0 + d)], \qquad (6.5.15)$$
$$A_{mn}^{(1)\pm} = A_{mn}^{(2)\pm} = A_{mn}^{(3)-} = B_{mn}^{\pm} = C_{mn}^{(1)\pm} = C_{mn}^{(2)\pm} = 0.$$

For numerical results, consider a plate with $a = 0.12\,\text{cm}$, $b = 0.1\,\text{cm}$, $h = 0.01\,\text{cm}$, $c = 0.015\,\text{cm}$, $d = 0.0125\,\text{cm}$, $x_0 = 0.06\,\text{cm}$, $y_0 = 0.05\,\text{cm}$, and $H^0 = 100\,\text{Oe}$. $p = -10^8\,\text{dyn/cm}^2$ is used under which m_1 and m_2 are much smaller than M^0. Equation (6.5.15)$_1$ shows that the decay rate of $A_{mn}^{(3)+}$ is

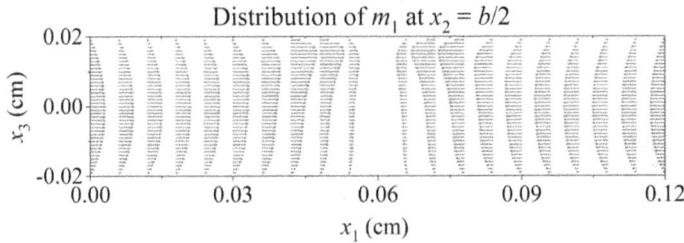

Fig. 6.9. Distributions of m_1 at $x_2 = b/2$.

proportional to $1/(mn)$. Therefore, in the numerical calculation, the terms of the trigonometric series are summed from one to an upper limit of mn. Numerical tests show that the series converges quickly enough for sufficient accuracy without encountering any numerical problems.

Figure 6.8 shows the distribution of $\mathbf{m}$ at the plate middle plane where $x_3 = 0$. $\mathbf{m}$ is relatively large near the central loading area and decays away from the center. Right at the center, because of symmetry, $\mathbf{m} = 0$.

Since the mechanical load p is applied to the plate top surface only and the bottom surface is free, we are also interested in how $\mathbf{m}$ varies along the plate thickness. Figure 6.9 shows $m_1(x_3)$ for different x_1 at a cross section of the plate where $x_2 = b/2$. Near the loading area, m_1 is relatively large, especially near the top surface. Far away from the loading area m_1 is smaller and is nearly symmetric about the plate middle plane. The distribution of $\mathbf{m}$ is related to the shear strain/stress distribution in the plate as shown in Eq. (6.2.10).

Chapter 7

ANTIPLANE PROBLEMS OF FERROMAGNETOELASTICITY

Antiplane or shear-horizontal (SH) motions of elastic crystals with electromagnetic couplings are relatively simple mathematically. These simple motions exist in certain crystals with relatively high symmetry. They can be used to exhibit the basic physics of mechanical, electric and magnetic interactions [35]. This chapter is on antiplane problems of saturated ferromagnetoelastic solids. It is based on the equations for small fields on a finite bias [18] in Chapter 6 for yttrium iron garnet or $Y_3Fe_5O_{12}$ (YIG), a cubic crystal of class (m3m). Gaussian units are used.

7.1 Governing Equations

Consider YIG with an initial magnetization $\mathbf{M}^0$ along the x_3 axis. The initial magnetic field is $\mathbf{H}^0$. For antiplane problems, we have $u_1 = u_2 = 0$ and $\partial/\partial x_3 = 0$. In this case Eqs. (6.2.13)–(6.2.18) reduce to the following four equations for the mechanical displacement u_3, magnetic potential ψ, m_1 and m_2 which are the two nonzero components of the incremental magnetization vector:

$$c_{44}(u_{3,11} + u_{3,22}) + 2b_{44}M^0(m_{1,1} + m_{2,2}) + f_3 = \rho\ddot{u}_3, \quad (7.1.1)$$

$$\psi_{,11} + \psi_{,22} - 4\pi(m_{1,1} + m_{2,2}) = 0, \quad (7.1.2)$$

$$M^0\psi_{,2} - 2\alpha_{11}M^0(m_{2,11} + m_{2,22}) + \chi(M^0)^3 m_2$$
$$+ 2b_{44}(M^0)^2 u_{3,2} + H^0 m_2 = \frac{1}{\gamma}\dot{m}_1, \quad (7.1.3)$$

$$-M^0\psi_{,1} + 2\alpha_{11}M^0(m_{1,11} + m_{1,22}) - \chi(M^0)^3 m_1$$
$$- 2b_{44}(M^0)^2 u_{3,1} - H^0 m_1 = \frac{1}{\gamma}\dot{m}_2. \quad (7.1.4)$$

$m_3 = 0$ because of the saturation condition that $\mathbf{M}^0 \cdot \mathbf{m} = 0$, where $\mathbf{M}^0$ has a fixed saturation magnitude. From Eqs. (6.2.10)–(6.2.12), the relevant constitutive relations for the incremental stress tensor $\boldsymbol{\tau}$, local magnetic field $\mathbf{h}^L$ and exchange tensor $\mathbf{a}$ are

$$\tau_{13} = c_{44}u_{3,1} + 2b_{44}M^0 m_1,$$
$$\tau_{23} = c_{44}u_{3,2} + 2b_{44}M^0 m_2, \tag{7.1.5}$$

$$h_1^L = -\chi(M^0)^2 m_1 - 2b_{44}M^0 u_{3,1},$$
$$h_2^L = -\chi(M^0)^2 m_2 - 2b_{44}M^0 u_{3,2}, \tag{7.1.6}$$

$$a_{11} = -2\alpha_{11}m_{1,1}, \quad a_{21} = -2\alpha_{11}m_{1,2},$$
$$a_{12} = -2\alpha_{11}m_{2,1}, \quad a_{22} = -2\alpha_{11}m_{2,2}. \tag{7.1.7}$$

Although YIG is nonpiezomagnetic in its natural state free from any fields, Eqs. (7.1.5) and (7.1.6) show that YIG is effectively piezomagnetic under $\mathbf{M}^0$ through the magnetostrictive constant b_{44}. Values of the effective material constants of YIG under $\mathbf{M}^0$ in Eqs. (7.1.5)–(7.1.7) can be found in Eqs. (6.2.6)–(6.2.8).

7.2   Coordinate-Independent Equations

We introduce a two-dimensional index notation that subscripts a and b assume 1 and 2 only without 3, and repeated indices are summed from 1 to 2. Then Eqs. (7.1.3) and (7.1.4) can be written as

$$M^0\psi_{,2} - 2\alpha_{11}M^0 m_{2,aa} + \chi(M^0)^3 m_2$$
$$+ 2b_{44}(M^0)^2 u_{3,2} + H^0 m_2 = \frac{1}{\gamma}\dot{m}_1, \tag{7.2.1}$$

$$-M^0\psi_{,1} + 2\alpha_{11}M^0 m_{1,aa} - \chi(M^0)^3 m_1$$
$$- 2b_{44}(M^0)^2 u_{3,1} - H^0 m_1 = \frac{1}{\gamma}\dot{m}_2. \tag{7.2.2}$$

We differentiate Eq. (7.2.1) with respect to x_2 and Eq. (7.2.2) with respect to x_1, and subtract the resulting equations to obtain

$$M^0\psi_{,aa} - 2\alpha_{11}M^0 m_{b,baa} + \chi(M^0)^3 m_{a,a}$$
$$+ 2b_{44}(M^0)^2 u_{3,aa} + H^0 m_{a,a} = \frac{1}{\gamma}(\dot{m}_{1,2} - \dot{m}_{2,1}). \tag{7.2.3}$$

We also differentiate Eq. (7.2.1) with respect to x_1 and Eq. (7.2.2) with respect to x_2, and add the resulting equations:

$$2\alpha_{11}M^0(m_{1,2} - m_{2,1})_{,aa} - \chi(M^0)^3(m_{1,2} - m_{2,1})$$
$$- H^0(m_{1,2} - m_{2,1}) = \frac{1}{\gamma}\dot{m}_{a,a}. \tag{7.2.4}$$

Let

$$\nabla = \mathbf{e}_1\partial_1 + \mathbf{e}_2\partial_2, \quad \nabla^2 = \partial_1^2 + \partial_2^2,$$
$$\mathbf{m} = m_1\mathbf{e}_1 + m_2\mathbf{e}_2, \tag{7.2.5}$$

and introduce Π and Ω through

$$\Pi = m_{a,a} = m_{1,1} + m_{2,2} = \nabla \cdot \mathbf{m},$$
$$\mathbf{\Omega} = \Omega\mathbf{e}_3 = (m_{2,1} - m_{1,2})\mathbf{e}_3 = \nabla \times \mathbf{m}. \tag{7.2.6}$$

Then Eqs. (7.1.1), (7.1.2), (7.2.3) and (7.2.4) can be written as four equations for u_3, ψ, Π and Ω as [36]:

$$c_{44}\nabla^2 u_3 + 2b_{44}M^0\Pi + f_3 = \rho\ddot{u}_3, \tag{7.2.7}$$

$$\nabla^2\psi - 4\pi\Pi = 0, \tag{7.2.8}$$

$$-2\alpha_{11}M^0\nabla^2\Pi + [\chi(M^0)^3 + H^0]\Pi$$
$$+ M^0\nabla^2\psi + 2b_{44}(M^0)^2\nabla^2 u_3 = -\frac{1}{\gamma}\dot{\Omega}, \tag{7.2.9}$$

$$-2\alpha_{11}M^0\nabla^2\Omega + [\chi(M^0)^3 + H^0]\Omega = \frac{1}{\gamma}\dot{\Pi}. \tag{7.2.10}$$

Equations (7.2.7)–(7.2.10) are coordinate-independent and can be broken into components in any coordinate systems conveniently.

For statics, Eqs. (7.2.7)–(7.2.10) reduce to

$$c_{44}\nabla^2 u_3 + 2b_{44}M^0\Pi + f_3 = 0, \tag{7.2.11}$$

$$\nabla^2\psi - 4\pi\Pi = 0, \tag{7.2.12}$$

$$-2\alpha_{11}M^0\nabla^2\Pi + [\chi(M^0)^3 + H^0]\Pi$$
$$+ M^0\nabla^2\psi + 2b_{44}(M^0)^2\nabla^2 u_3 = 0, \tag{7.2.13}$$

$$-2\alpha_{11}M^0\nabla^2\Omega + [\chi(M^0)^3 + H^0]\Omega = 0. \tag{7.2.14}$$

Equation (7.2.14) can be written as

$$\nabla^2\Omega = \lambda^2\Omega, \tag{7.2.15}$$

where

$$\lambda^2 = \frac{\chi(M^0)^3 + H^0}{2\alpha_{11}M^0}.$$ (7.2.16)

From Eqs. (7.2.11)–(7.2.13), an equation for Π similar to Eq. (7.2.15) can be obtained. In one-dimensional antiplane problems depending on x_1 only, Eq. (7.2.15) reduces to

$$\Omega_{,11} = \lambda^2\Omega$$ (7.2.17)

whose general solution is

$$\Omega = C_1 \exp(\lambda x_1) + C_2 \exp(-\lambda x_1).$$ (7.2.18)

Equation (7.2.18) shows the basic behavior of static distributions of spin fields. The fields are exponentially growing or decaying. In polar coordinates, for axisymmetric problems, Eq. (7.2.15) takes the following form:

$$\nabla^2\Omega = \frac{1}{r}\frac{\partial}{\partial r}\left(r\frac{\partial\Omega}{\partial r}\right) = \frac{1}{r}\left(\frac{\partial\Omega}{\partial r} + r\frac{\partial^2\Omega}{\partial r^2}\right) = \frac{\partial^2\Omega}{\partial r^2} + \frac{1}{r}\frac{\partial\Omega}{\partial r} = \lambda^2\Omega,$$ (7.2.19)

which is the modified Bessel equation of order zero. Its general solution is

$$\Omega = C_1 I_0(\lambda r) + C_2 K_0(\lambda r),$$ (7.2.20)

where I_0 and K_0 are the zero-order modified Bessel functions of the first and second kinds, respectively. $I_0(0) = 1$ and it grows unboundedly with r. $K_0(0) = \infty$ and it decays with r.

7.3   Effects of Local Mechanical Loads

In this section, we analyze a static antiplane problem over a finite domain [36]. Consider a rectangular domain within $0 < x_1 < a$ and $0 < x_2 < b$ as shown in Fig. 7.1. f_3 is uniform within a small rectangular area centered at (x_0, y_0). Outside shaded rectangle we have $f_3 = 0$. We are mainly interested in the effect of f_3 on the distributions of m_1 and m_2, particularly near the loading area. For statics, Eqs. (7.1.1)–(7.1.4) reduce to

$$c_{44}(u_{3,11} + u_{3,22}) + 2b_{44}M^0(m_{1,1} + m_{2,2}) + f_3 = 0,$$ (7.3.1)

$$\psi_{,11} + \psi_{,22} - 4\pi(m_{1,1} + m_{2,2}) = 0,$$ (7.3.2)

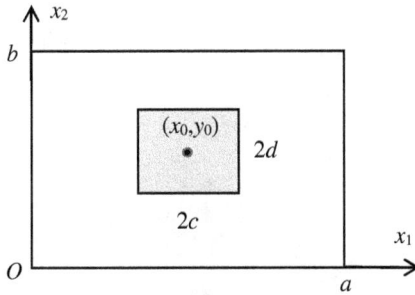

Fig. 7.1. A rectangular domain and a local loading area centered at (x_0, y_0).

$$M^0\psi_{,2} - 2\alpha_{11}M^0(m_{2,11} + m_{2,22}) + \chi(M^0)^3 m_2$$
$$+ 2b_{44}(M^0)^2 u_{3,2} + H^0 m_2 = 0, \tag{7.3.3}$$

$$-M^0\psi_{,1} + 2\alpha_{11}M^0(m_{1,11} + m_{1,22}) - \chi(M^0)^3 m_1$$
$$- 2b_{44}(M^0)^2 u_{3,1} - H^0 m_1 = 0. \tag{7.3.4}$$

The boundaries of the rectangular domain are fixed mechanically and have a constant magnetic potential ($\psi = 0$) which is known as a perfect magnetic wall or perfect magnetic conductor boundary. The boundary conditions are

$$
\begin{aligned}
u_3 = 0, \quad \psi = 0, \quad m_{1,1} = 0, \quad m_2 = 0, \quad x_1 = 0, a, \\
u_3 = 0, \quad \psi = 0, \quad m_1 = 0, \quad m_{2,2} = 0, \quad x_2 = 0, b,
\end{aligned}
\tag{7.3.5}
$$

where the boundary conditions on **m** are mixed in the sense that **m** or its normal derivatives are prescribed. For a solution to Eqs. (7.3.1)–(7.3.5), we let

$$\{f_3, u_3, \psi\} = \sum_{m=1}^{\infty}\sum_{n=1}^{\infty}\{F_{mn}, U_{mn}, \Psi_{mn}\}\sin(\xi_m x_1)\sin(\eta_n x_2), \tag{7.3.6}$$

$$m_1 = \sum_{m=1}^{\infty}\sum_{n=1}^{\infty} V_{mn}\cos(\xi_m x_1)\sin(\eta_n x_2),$$

$$m_2 = \sum_{m=1}^{\infty}\sum_{n=1}^{\infty} W_{mn}\sin(\xi_m x_1)\cos(\eta_n x_2), \tag{7.3.7}$$

where

$$\xi_m = \frac{m\pi}{a}, \quad \eta_n = \frac{n\pi}{b}. \tag{7.3.8}$$

(a)

(b)

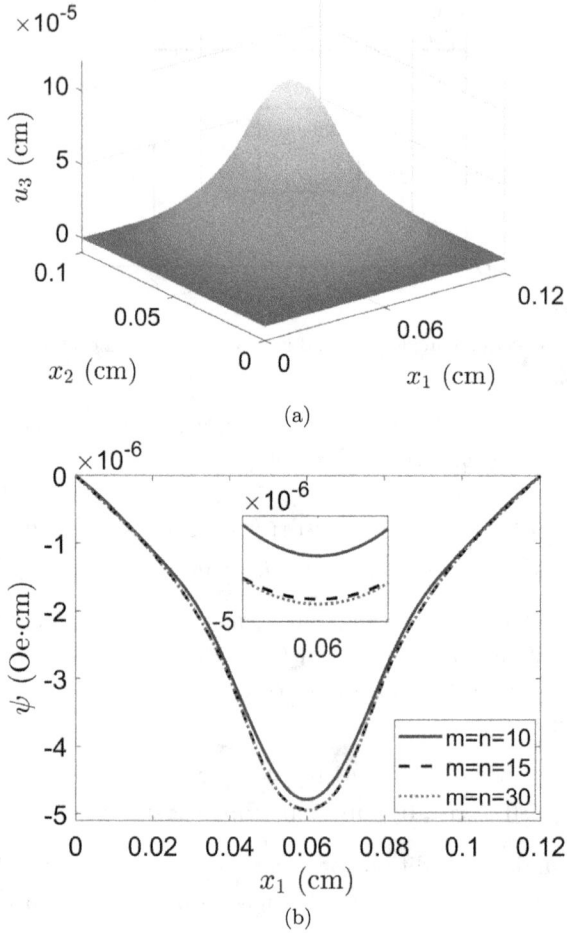

Fig. 7.2. (a) Distribution of u_3. (b) Convergence of ψ along $x_2 = y_0$.

U_{mn}, Ψ_{mn}, V_{mn} and W_{mn} are undetermined constants. F_{mn} are known from the applied mechanical load. Denoting $f_3 = f$ in the local loading area, we have

$$F_{mn} = \frac{4}{ab} \frac{f}{\xi_m \eta_n} [\cos(\xi_m(x_0 - c)) - \cos(\xi_m(x_0 + c))]$$
$$\times [\cos(\eta_n(y_0 - d)) - \cos(\eta_n(y_0 + d))]. \qquad (7.3.9)$$

It can be verified that Eqs. (7.3.6) and (7.3.7) satisfy Eq. (7.3.5). Substituting Eqs. (7.3.6) and (7.3.7) into Eqs. (7.3.1)–(7.3.4), we obtain a system of

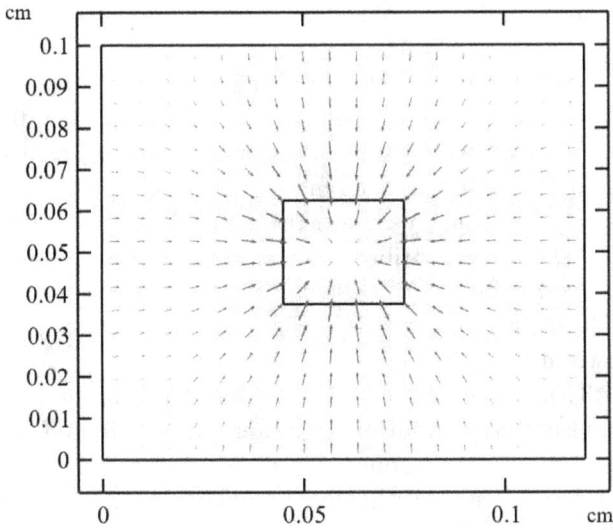

(a)

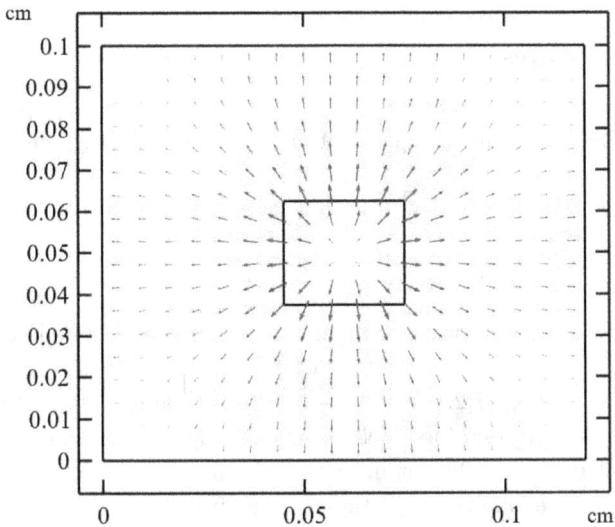

(b)

Fig. 7.3. Distribution of $\mathbf{m} = m_1\mathbf{e}_1 + m_2\mathbf{e}_2$. (a) $f = 4 \times 10^{11}\,\mathrm{dyn/cm}^3$. (b) $f = -4 \times 10^{11}\,\mathrm{dyn/cm}^3$.

linear algebraic equations for the undetermined constants, which is solved on a computer using MATLAB.

For numerical results, consider a rectangular domain with $a = 0.12\,\text{cm}$, $b = 0.1\,\text{cm}$, $c = 0.015\,\text{cm}$, $d = 0.0125\,\text{cm}$, $x_0 = 0.06\,\text{cm}$, $y_0 = 0.05\,\text{cm}$ and $H_0 = 100\,\text{Oe}$. $f = 4 \times 10^{11}\,\text{dyn/cm}^3$ for which m_1 and m_2 are significantly smaller than M^0. The summation range of (m, n) is $(30, 30)$. Some of these parameters may be varied in some of the figures below.

Figure 7.2 shows the distributions of u_3 and ψ. Under a local load near the center, the distribution of u_3 is as simple as what is shown in (a) as expected. The distribution of ψ is similar. The series converges rapidly as what is shown in (b).

Figure 7.3 shows the effects of f on the distribution of $\mathbf{m}$. It can be seen that a local mechanical load near the center has a significant effect on the distribution of $\mathbf{m}$ near the loading area. $\mathbf{m}$ decays rapidly from the loading area. This is consistent with the decaying behavior of the static fields of $\mathbf{m}$ observed in Section 7.2. When the mechanical load is reversed, so is $\mathbf{m}$. This is as expected from a linear theory.

7.4   Plane Waves

Consider the propagation of the following plane waves governed by Eqs. (7.2.7)–(7.2.10) in the x_1 direction of an unbounded two-dimensional domain of (x_1, x_2):

$$
\begin{aligned}
u_3 &= U \sin(\xi x_1 - \omega t), \quad \psi = \Psi \sin(\xi x_1 - \omega t), \\
\Pi &= V \sin(\xi x_1 - \omega t), \quad \Omega = W \cos(\xi x_1 - \omega t),
\end{aligned}
\tag{7.4.1}
$$

where ξ and ω are the wave number and frequency. U, Ψ, V and W are wave amplitudes. The substitution of Eq. (7.4.1) into the homogeneous form of Eqs. (7.2.7)–(7.2.10) with $f_3 = 0$ results in a system of linear and homogeneous equations for U, Ψ, V and W. For nontrivial solutions the determinant of the coefficient matrix of the linear equations must vanish, which leads to the following equation that determines the dispersion relation between ξ and ω [18]:

$$
\begin{aligned}
&\alpha^2 c_{44} \xi^6 - [\alpha^2 \rho \omega^2 - c_{44}\alpha(2P + 4\pi) + e^2\alpha]\xi^4 \\
&- \left[e^2 P - c_{44}P(P + 4\pi) + \omega^2 \left(\frac{c_{44}}{\gamma^2(M^0)^2} + 2\rho\alpha P + 4\pi\rho\alpha \right) \right] \xi^2 \\
&\quad - \rho P(P + 4\pi)\omega^2 + \frac{\rho}{\gamma^2(M^0)^2}\omega^4 = 0,
\end{aligned}
\tag{7.4.2}
$$

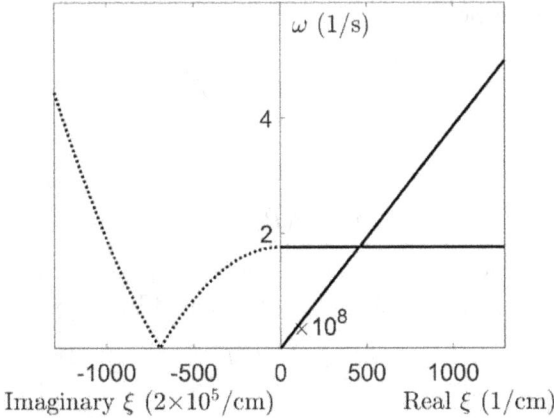

Fig. 7.4. Dispersion curves of coupled elastic and spin waves when $b_{44} \neq 0$.

where we have denoted

$$e = 2b_{44}M^0, \quad \alpha = 2\alpha_{11},$$
$$P = H^0/M^0 + K, \quad K = \chi(M^0)^2. \tag{7.4.3}$$

The dispersion curves determined from Eq. (7.4.2) are shown in Fig. 7.4.

In the special case of $b_{44} = 0$, we have $e = 0$. The coupling between elastic and spin waves disappears. In this case, the dispersion relations separate into a branch for uncoupled elastic waves with:

$$\omega^2 = \frac{c_{44}}{\rho}\xi^2, \quad \frac{\omega}{\xi} = \sqrt{\frac{c_{44}}{\rho}}, \tag{7.4.4}$$

and branches for uncoupled spin waves with:

$$\frac{\omega^2}{\gamma^2(M^0)^2} = \alpha^2\xi^4 + \alpha(2P + 4\pi)\xi^2 + P(P + 4\pi). \tag{7.4.5}$$

When $\xi = 0$, the uncoupled spin waves have the following cutoff frequency:

$$\omega|_{\xi=0} = \gamma M^0\sqrt{P(P + 4\pi)} = \omega_c. \tag{7.4.6}$$

When $b_{44} \neq 0$, for long waves with a small ξ, Eq. (7.4.2) may be approximated by

$$-\left[e^2 P - c_{44}P(P + 4\pi) + \omega^2\left(\frac{c_{44}}{\gamma^2(M^0)^2} + 2\rho\alpha P + 4\pi\rho\alpha\right)\right]\xi^2$$
$$-\rho P(P + 4\pi)\omega^2 + \frac{\rho}{\gamma^2(M^0)^2}\omega^4 = 0. \tag{7.4.7}$$

The cutoff frequencies of Eq. (7.4.7) are determined by

$$-\rho P(P + 4\pi)\omega^2 + \frac{\rho}{\gamma^2 (M^0)^2}\omega^4 = 0. \tag{7.4.8}$$

Its roots are

$$\omega = 0,$$
$$\omega = \gamma M^0 \sqrt{P(P + 4\pi)} = \omega_c, \tag{7.4.9}$$

which are the same as the ones predicted by Eq. (7.4.4) and (7.4.6). Near $(\xi, \omega) = (0, 0)$, Eq. (7.4.7) may be approximated by

$$[e^2 P - c_{44} P(P + 4\pi)]\xi^2 + \rho P(P + 4\pi)\omega^2 = 0, \tag{7.4.10}$$

which determines

$$\frac{\omega}{\xi} = \sqrt{\frac{c_{44}(P + 4\pi) - e^2}{\rho(P + 4\pi)}}. \tag{7.4.11}$$

Equation (7.4.11) describes an elastic shear wave with magnetic coupling. In the special case of $e = 0$, it reduces to the purely elastic wave in Eq. (7.4.4). Near $(\xi, \omega) = (0, \omega_c)$, let

$$\omega^2 = \omega_c^2 + \Delta. \tag{7.4.12}$$

Then Eq. (7.4.7) becomes

$$-\left[e^2 P - c_{44} P(P + 4\pi) + \omega_c^2 \left(\frac{c_{44}}{\gamma^2 (M^0)^2} + 2\rho\alpha P + 4\pi\rho\alpha \right) \right] \xi^2$$
$$-\rho P(P + 4\pi)(\omega_c^2 + \Delta) + \frac{\rho}{\gamma^2 (M^0)^2}(\omega_c^2 + \Delta)^2 = 0. \tag{7.4.13}$$

For small Δ, Eq. (7.4.13) may be approximated by

$$-\left[e^2 P - c_{44} P(P + 4\pi) + \omega_c^2 \frac{c_{44}}{\gamma^2 (M^0)^2} + \omega_c^2 (2\rho\alpha P + 4\pi\rho\alpha) \right] \xi^2$$
$$-\rho P(P + 4\pi)(\omega_c^2 + \Delta) + \frac{\rho}{\gamma^2 (M^0)^2}(\omega_c^4 + 2\omega_c^2\Delta) = 0. \tag{7.4.14}$$

Equation (7.4.14) leads to

$$\Delta = \frac{e^2 P + \omega_c^2 2\alpha\rho(P + 2\pi)}{\rho P(P + 4\pi)}\xi^2. \tag{7.4.15}$$

Substituting Eq. (7.4.15) into Eq. (7.4.12), we obtain

$$\omega^2 = \omega_c^2 + \frac{e^2 P + \omega_c^2 2\alpha\rho(P + 2\pi)}{\rho P(P + 4\pi)}\xi^2. \tag{7.4.16}$$

In the special case of $e = 0$, Eq. (7.4.16) reduces to

$$\frac{\omega^2}{(\gamma M^0)^2} = P(P + 4\pi) + 2\alpha(P + 2\pi)\xi^2, \qquad (7.4.17)$$

which agrees with Eq. (7.4.5) for small ξ.

7.5 Vibration of a Rectangular Body

In this section, we study the free vibration of the rectangular body shown in Fig. 7.1 [36] with $f_3 = 0$. The body has a mechanically fixed boundary which is a perfect magnetic wall. The boundary conditions are taken as

$$u_3 = 0, \quad \psi = 0, \quad \Pi = 0, \quad \Omega = 0. \qquad (7.5.1)$$

For a solution of Eqs. (7.2.7)–(7.2.10) and (7.5.1), let

$$\{u_3, \psi, \Pi\} = \{U_{mn}, \Psi_{mn}, V_{mn}\} \sin(\xi_m x_1) \sin(\eta_n x_2) \sin(\omega t), \quad (7.5.2)$$

$$\Omega = W_{mn} \sin(\xi_m x_1) \sin(\eta_n x_2) \cos(\omega t), \qquad (7.5.3)$$

$$\xi_m = \frac{m\pi}{a}, \quad \eta_n = \frac{n\pi}{b}, \quad m, n = 1, 2, 3, \ldots. \qquad (7.5.4)$$

Equations (7.5.2) and (7.5.3) satisfy Eq. (7.5.1). Substituting Eqs. (7.5.2) and (7.5.3) into Eqs. (7.2.7)–(7.2.10) with $f_3 = 0$, we obtain a system of four linear and homogeneous equations for U_{mn}, Ψ_{mn}, V_{mn} and W_{mn} for a fixed pair of (m, n). For nontrivial solutions the determinant of the coefficient matrix of the equations must vanish, which leads to the following frequency equation:

$$\begin{aligned}
\alpha^2 & c_{44}(\xi_m^2 + \eta_n^2)^3 \\
& - [\alpha^2 \rho \omega^2 - c_{44}\alpha(2P + 4\pi) + e^2\alpha](\xi_m^2 + \eta_n^2)^2 \\
& - \left[e^2 P - c_{44}P(P + 4\pi) + \omega^2 \left(\frac{c_{44}}{\gamma^2(M^0)^2} + 2\rho\alpha P + 4\pi\rho\alpha \right) \right] \\
& \times (\xi_m^2 + \eta_n^2) - \rho P(P + 4\pi)\omega^2 + \frac{\rho}{\gamma^2(M^0)^2}\omega^4 = 0,
\end{aligned} \qquad (7.5.5)$$

where

$$\begin{aligned}
e &= 2b_{44}M^0, \quad \alpha = 2\alpha_{11}, \\
P &= H^0/M^0 + K, \quad K = \chi(M^0)^2.
\end{aligned} \qquad (7.5.6)$$

In the special case of $b_{44} = 0$, the coupling between elastic and magnetic modes disappears. In this case, the frequency equation in Eq. (7.5.5) leads to the following frequencies of uncoupled elastic modes with

$$\omega^2 = c_{44}(\xi_m^2 + \eta_n^2)/\rho, \tag{7.5.7}$$

and uncoupled magnetic modes with

$$\begin{aligned}
&\omega^2/[\gamma^2(M^0)^2] \\
&= \alpha^2(\xi_m^2 + \eta_n^2)^2 + \alpha(2P + 4\pi)(\xi_m^2 + \eta_n^2) + P(P + 4\pi).
\end{aligned} \tag{7.5.8}$$

When $b_{44} \neq 0$, the frequencies in Eqs. (7.5.7) and (7.5.8) change a little duo to magnetoelastic coupling. We will call the corresponding modes essentially elastic or essentially magnetic modes. Once ω is known from Eq. (7.5.5), the corresponding U_{mn}, Ψ_{mn}, V_{mn} and W_{mn} are determined from the linear equations governing them. U_{mn} is normalized to one. Then **m** is determined from Eq. (7.2.6), i.e., from

$$\Pi = m_{1,1} + m_{2,2}, \quad \Omega = m_{2,1} - m_{1,2}. \tag{7.5.9}$$

The result is

$$\begin{aligned}
m_1 &= -\frac{\xi_{mn} V_{mn} \cos(\xi_{mn} x_1) \sin(\eta_{mn} x_2) \sin(\omega t)}{\xi_{mn}^2 + \eta_{mn}^2} \\
&+ \frac{\eta_{mn} W_{mn} \sin(\xi_{mn} x_1) \cos(\eta_{mn} x_2) \cos(\omega t)}{\xi_{mn}^2 + \eta_{mn}^2},
\end{aligned} \tag{7.5.10}$$

$$\begin{aligned}
m_2 &= -\frac{\eta_{mn} V_{mn} \sin(\xi_{mn} x_1) \cos(\eta_{mn} x_2) \sin(\omega t)}{\xi_{mn}^2 + \eta_{mn}^2} \\
&- \frac{\xi_{mn} W_{mn} \cos(\xi_{mn} x_1) \sin(\eta_{mn} x_2) \cos(\omega t)}{\xi_{mn}^2 + \eta_{mn}^2}.
\end{aligned} \tag{7.5.11}$$

Figure 7.5 shows the distribution of **m** of the first essentially elastic mode with $(m, n) = (1, 1)$ and $\omega = 1.5717 \times 10^7 \, (1/\text{s})$ at two time instants with $\omega t = 0$ and $\pi/4$. **m** is periodic in ωt with a period of 2π. Figure 7.5 shows that elastic deformations are accompanied by distributions of **m** due to the effective piezomagnetic coupling. When $(m, n) = (1, 1)$, the displacement gradient is small near the center. Hence, the effective piezomagnetic coupling is weak near the center with a vanishing **m** at the center.

Figure 7.6 shows the distribution of **m** of two other essentially elastic modes when $t = 0$. As (m, n) increases, the modes have more variations along x_1 and/or x_2. These are typical behaviors of higher-order modes.

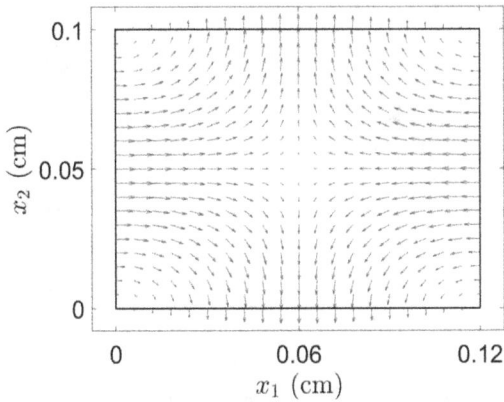

(a)

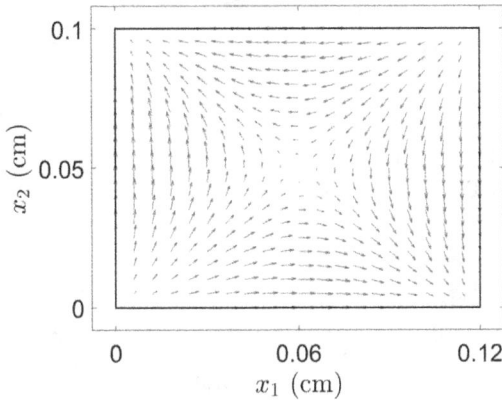

(b)

Fig. 7.5. Distribution of **m** of the first essentially elastic mode with $(m, n) = (1, 1)$ and $\omega = 1.5717 \times 10^7$ (1/s). (a) $\omega t = 0$. (b) $\omega t = \pi/4$.

The distributions of **m** of the essentially magnetic modes are qualitatively similar and are not presented. Their frequencies are almost the same for the first few modes. This is because the expression of the frequencies of the essentially magnetic modes has a constant term independent of ξ_m and η_n. This constant term determines the cutoff frequency. When (m, n) is small, the constant term dominates. In this case, the frequencies are not sensitive to (m, n) and are close to the cutoff frequency.

(a)

(b)

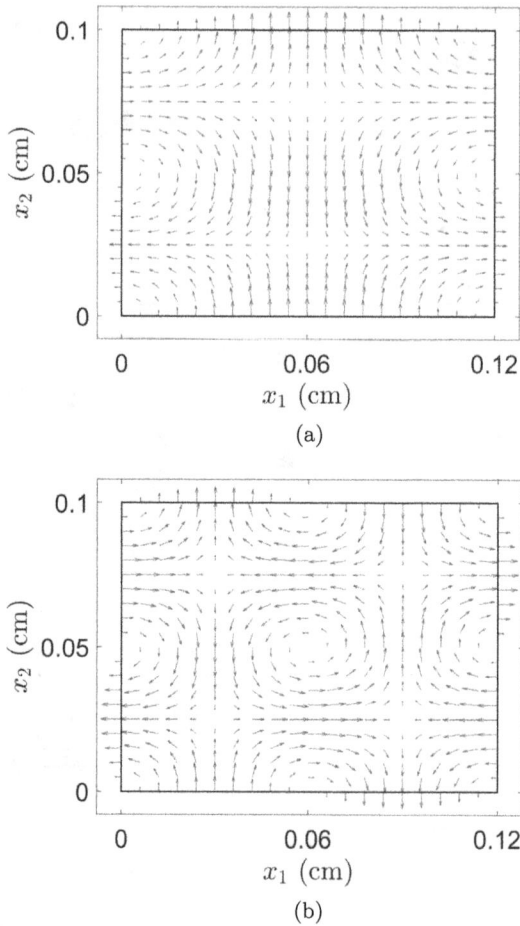

Fig. 7.6. Distribution of **m** of two essentially elastic modes when $t = 0$. (a) $(m, n) = (2, 1)$ and $\omega = 2.3469 \times 10^7$ (1/s). (b) $(m, n) = (2, 2)$ and $\omega = 3.1435 \times 10^7$ (1/s).

7.6   Coupling to Electromagnetic Waves

As motions of magnetic moments, spin waves interact with electromagnetic waves directly through Maxwell's equations which is referred to as magnon–photon coupling. Thus, in deformable ferromagnetic solids, acoustic waves and electromagnetic waves can interact indirectly through spin waves. Let $\mathbf{e}^M$, $\mathbf{d}^M$, $\mathbf{b}^M$ and $\mathbf{h}^M$ be the small and incremental Maxwellian electric field, electric displacement, magnetic induction and magnetic field on top of the initial and static fields associated with $\mathbf{M}^0$. They are governed by

Maxwell's equations

$$\nabla \cdot \mathbf{d}^M = 0, \tag{7.6.1}$$

$$\nabla \cdot \mathbf{b}^M = 0, \tag{7.6.2}$$

$$\nabla \times \mathbf{e}^M + \frac{1}{c}\frac{\partial \mathbf{b}^M}{\partial t} = 0, \tag{7.6.3}$$

$$\nabla \times \mathbf{h}^M = \frac{1}{c}\frac{\partial \mathbf{d}^M}{\partial t}, \tag{7.6.4}$$

where c is the speed of light in a vacuum. Equations (7.6.1) and (7.6.2) are essentially implied by Eqs. (7.6.3) and (7.6.4). For simplicity, electromechanical and magnetoelectric couplings are not considered. We have the following relationships:

$$\begin{aligned} d_i^M &= e_i^M + 4\pi p_i, \\ b_i^M &= h_i^M + 4\pi m_i - 4\pi M_i^0 u_{j,j}, \end{aligned} \tag{7.6.5}$$

where $\mathbf{p}$ is the incremental electric polarization vector. For YIG we have

$$d_i^M = \varepsilon e_i^M, \tag{7.6.6}$$

where ε is the dielectric constant of the material.

Consider antiplane problems with $u_1 = u_2 = 0$ and $\partial/\partial x_3 = 0$. The relevant electric fields are [37]

$$\begin{aligned} e_1^M &= 0, \quad e_2^M = 0, \quad e_3^M = e_3^M(x_1, x_2, t), \\ d_1^M &= 0, \quad d_2^M = 0, \quad d_3^M = e_3^M(x_1, x_2, t). \end{aligned} \tag{7.6.7}$$

In this case, Eqs. (7.6.3) and (7.6.4) reduce to

$$e_{3,2}^M + \frac{1}{c}\frac{\partial b_1^M}{\partial t} = 0, \ -e_{3,1}^M + \frac{1}{c}\frac{\partial b_2^M}{\partial t} = 0, \tag{7.6.8}$$

$$h_{2,1}^M - h_{1,2}^M = \frac{1}{c}\frac{\partial d_3^M}{\partial t}. \tag{7.6.9}$$

We also have

$$\begin{aligned} c_{44}(u_{3,11} + u_{3,22}) + 2b_{44}M^0(m_{1,1} + m_{2,2}) &= \rho \ddot{u}_3, \\ -M^0 h_2^M - 2\alpha_{11}M^0(m_{2,11} + m_{2,22}) & \\ +\chi(M^0)^3 m_2 + 2b_{44}(M^0)^2 u_{3,2} + H^0 m_2 &= \frac{1}{\gamma}\dot{m}_1, \\ M^0 h_1^M + 2\alpha_{11}M^0(m_{1,11} + m_{1,22}) & \\ -\chi(M^0)^3 m_1 - 2b_{44}(M^0)^2 u_{3,1} - H^0 m_1 &= \frac{1}{\gamma}\dot{m}_2. \end{aligned} \tag{7.6.10}$$

Equations (7.6.8)–(7.6.10) can be written as six equations for u_3, e_3^M, h_1^M, h_2^M, m_1 and m_2.

Let

$$u_3 = A_1 \exp[i(\xi x_1 - \omega t)], \quad e_3^M = A_2 \exp[i(\xi x_1 - \omega t)],$$
$$h_1^M = A_3 \exp[i(\xi x_1 - \omega t)], \quad h_2^M = A_4 \exp[i(\xi x_1 - \omega t)], \quad (7.6.11)$$
$$m_1 = A_5 \exp[i(\xi x_1 - \omega t)], \quad m_2 = A_6 \exp[i(\xi x_1 - \omega t)].$$

The substitution of Eq. (7.6.11) into Eqs. (7.6.8)–(7.6.10) results in six linear and homogeneous equations for A_1 through A_6. For nontrivial solutions, the determinant of the coefficient matrix of the equations has to vanish. This leads to the following equation that determines the dispersion relations of the waves:

$$-c^2\xi^2 \left\{ \alpha^2 c_{44}\xi^6 - (\alpha^2\rho\omega^2 - c_{44}\alpha(2P + 4\pi) + \alpha e^2)\xi^4 \right.$$

$$- \left[e^2 P - c_{44}P(P + 4\pi) + \omega^2 \left(\frac{c_{44}}{\gamma^2(M^0)^2} + \rho\alpha(2P + 4\pi) \right) \right] \xi^2$$

$$\left. - \rho P(P + 4\pi)\omega^2 + \frac{\rho}{\gamma^2(M^0)^2}\omega^4 \right\}$$

$$+ \varepsilon\omega^2 \left\{ c_{44}\alpha^2\xi^6 - (\alpha^2\rho\omega^2 - 2c_{44}\alpha(P + 4\pi) + \alpha e^2)\xi^4 \right.$$

$$- \left[e^2(P + 4\pi) - c_{44}(P + 4\pi)^2 + \omega^2 \left(\frac{c_{44}}{\gamma^2(M^0)^2} + 2\rho\alpha(P + 4\pi) \right) \right] \xi^2$$

$$\left. - \rho(P + 4\pi)^2\omega^2 + \frac{\rho}{\gamma^2(M^0)^2}\omega^4 \right\} = 0, \quad (7.6.12)$$

where

$$e = 2b_{44}M^0, \quad \alpha = 2\alpha_{11},$$
$$P = H^0/M^0 + K, \quad K = \chi(M^0)^2. \quad (7.6.13)$$

In the special case of $b_{44} = 0$, the magnetoelastic coupling disappears and Eq. (7.6.12) reduces to a product of two factors. One is for uncoupled acoustic waves with

$$\rho\omega^2 - c_{44}\xi^2 = 0. \quad (7.6.14)$$

The other is for coupled electromagnetic and spin waves:

$$-c^2\xi^2 \left[\frac{\omega^2}{\gamma^2(M^0)^2} - (\alpha\xi^2 + P)(\alpha\xi^2 + P + 4\pi) \right]$$

$$+ \varepsilon\omega^2 \left[\frac{\omega^2}{\gamma^2(M^0)^2} - (\alpha\xi^2 + P + 4\pi)^2 \right] = 0. \quad (7.6.15)$$

When $c \to \infty$, Eq. (7.6.15) reduces to the following dispersion relation for uncoupled spin waves with

$$\frac{\omega^2}{\gamma^2 (M^0)^2} = \alpha^2 \xi^4 + \alpha(2P + 4\pi)\xi^2 + P(P + 4\pi). \qquad (7.6.16)$$

When $\alpha = 0$ and $\gamma \to \infty$, Eq. (7.6.15) reduces to the following dispersion relation for uncoupled electromagnetic waves:

$$\frac{\omega^2}{\xi^2} = \frac{c^2 P}{\varepsilon (P + 4\pi)}. \qquad (7.6.17)$$

When $\varepsilon = 1$ and $M^0 \to 0$, we have $P \to \infty$ and Eq. (7.6.17) reduces to $\omega/\xi = c$ for electromagnetic waves in a vacuum.

When $H^0 = 1500\,\text{Oe}$ and $\varepsilon = 14$, the dispersion relations of the uncoupled waves in Eqs. (7.6.14), (7.6.16) and (7.6.17) are shown in Fig. 7.7 in logarithmic scales for both the coordinate and the abscissa. The acoustic and electromagnetic waves are represented by straight lines and are nondispersive, with the electromagnetic waves at higher frequencies. The spin waves are represented by a curve. Each straight line intersects with the curve at two points.

When $H^0 = 1500\,\text{Oe}$ and $\varepsilon = 14$, the dispersion relations of the coupled waves determined by Eq. (7.6.12) are shown in Fig. 7.8. Near B and C there are couplings between acoustic and spin waves. Near A and D there

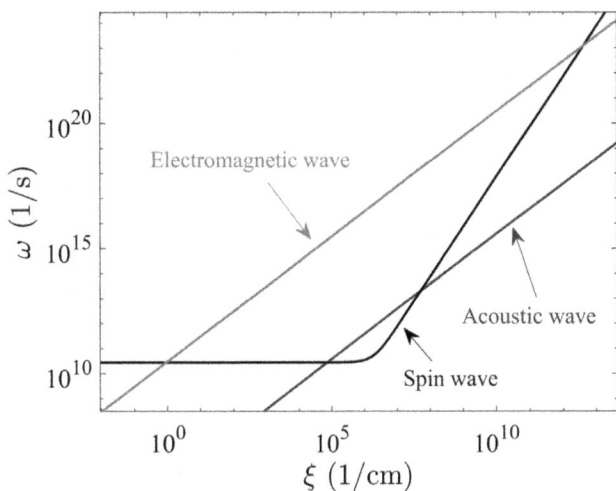

Fig. 7.7. Dispersion curves of uncoupled waves.

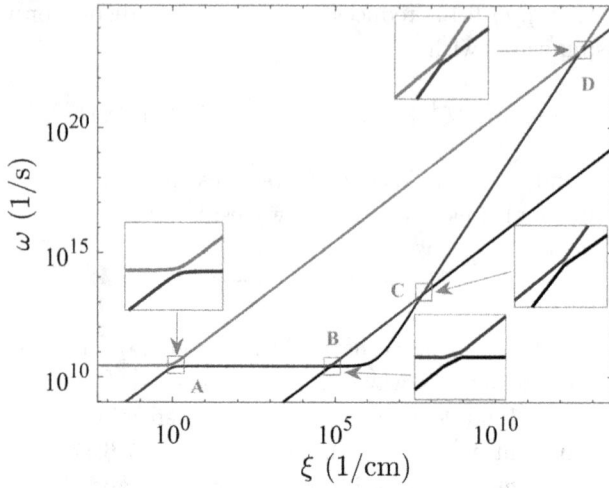

Fig. 7.8.   Dispersion curves of coupled waves.

are couplings between electromagnetic and spin waves. Since the material does not have electromechanical couplings, there is no direct interaction between acoustic and electromagnetic waves but they are related by the spin waves.

Chapter 8

Ferromagnetoelastic Plates

Thin films or plates are common structures in devices. Two-dimensional theories of plates are effective tools for modeling thin film devices. Plate theories can be derived from two-dimensional models directly or from three-dimensional theories by various approximations in the plate thickness coordinate [4, 38–41]. In this chapter, a set of two-dimensional equations for small fields superposed on a finite bias in saturated ferromagnetoelastic plates is derived using power series expansions in the plate thickness coordinate [42]. The equations are in the notation of [18] in Gaussian units.

8.1 Recapitulation of Three-Dimensional Equations

For convenience we collect the relevant three-dimensional equations in Section 6.1 below, from which two-dimensional plate equations will be derived systematically.

$$\tau_{ij,i} + f_j + f_j^M = \rho \ddot{u}_j, \tag{8.1.1}$$

$$h_i^M = -\psi_{,i}, \tag{8.1.2}$$

$$b_{i,i}^M = 0, \tag{8.1.3}$$

$$\varepsilon_{ijk} M_j^0 (h_k^M - a_{lk,l} + h_k^L) + \varepsilon_{ijk} m_j H_k^0 = \frac{1}{\gamma} \dot{m}_i, \tag{8.1.4}$$

$$f_j^M = M_i^0 h_{j,i}^M, \tag{8.1.5}$$

$$\tau_{ij} = \tau_{ij}^S + \tau_{ij}^A, \tag{8.1.6}$$

$$\tau_{ij}^A = \frac{1}{2}(M_i^0 h_j^L - M_j^0 h_i^L), \tag{8.1.7}$$

$$\tau_{ij}^S = \bar{c}_{ijkm} u_{k,m} + \bar{h}_{kij} m_k,$$

$$h_i^L = \bar{\chi}_{ik} m_k + \bar{g}_{ikm} u_{k,m}, \tag{8.1.8}$$

139

$$a_{ij} = \overline{\beta}_{ijkl} m_{l,k},$$

$$b_i^M = h_i^M + 4\pi m_i - 4\pi M_i^0 u_{j,j}. \tag{8.1.9}$$

With successive substitutions, Eqs. (8.1.1), (8.1.3) and (8.1.4) may be written as seven equations for the components of $\mathbf{u}$, $\mathbf{m}$ and ψ.

8.2   Hierarchy of Two-Dimensional Equations

Consider the thin plate with its normal along the x_3 axis as shown in Fig. 8.1. The x_1 and x_2 axes are in the middle plane of the plate. The in-plane normal and tangent at the plate boundary are $\mathbf{n}$ and $\mathbf{s}$. The plate thickness is $2c$. We begin with the following power series expansions of $\mathbf{u}$, $\mathbf{m}$ and ψ in the plate thickness coordinate x_3:

$$\{u_i, m_i, \psi\} = \sum_{n=0}^{\infty} \{u_i^{(n)}, m_i^{(n)}, \psi^{(n)}\} x_3^n. \tag{8.2.1}$$

Our goal is to obtain two-dimensional equations governing $u_i^{(n)}$, $m_i^{(n)}$ and $\psi^{(n)}$ which depend on x_1, x_2 and t. With Eq. (8.2.1), we write the gradients of $\mathbf{u}$, $\mathbf{m}$ and ψ as

$$\{u_{k,m}, m_{l,k}, h_i^M\} = \sum_{n=0}^{\infty} \{u_{km}^{(n)}, m_{lk}^{(n)}, h_i^{M(n)}\} x_3^n, \tag{8.2.2}$$

where

$$u_{km}^{(n)} = u_{k,m}^{(n)} + (n+1)\delta_{3m} u_k^{(n+1)},$$

$$m_{lk}^{(n)} = m_{l,k}^{(n)} + (n+1)\delta_{3k} m_l^{(n+1)}, \tag{8.2.3}$$

$$h_i^{M(n)} = -\psi_{,i}^{(n)} - (n+1)\delta_{3i} \psi^{(n+1)}.$$

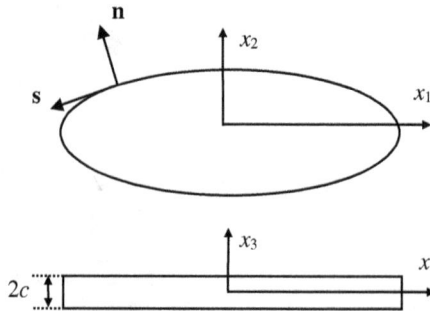

Fig. 8.1.   Plan view and cross section of a thin plate.

The symmetric parts of the plate extensional and shear forces, bending and twisting moments, magnetic and exchange resultants over a cross-section are defined by the following integrations along the plate thickness and are calculated using Eqs. (8.1.8), (8.2.1) and (8.2.2) as

$$\tau_{ij}^{S(m)} = \int_{-c}^{c} \tau_{ij}^{S} x_3^m dx_3 = \sum_{n=0}^{\infty} B^{(mn)}(\bar{c}_{ijkm} u_{km}^{(n)} + \bar{h}_{kij} m_k^{(n)}),$$

$$h_i^{L(m)} = \int_{-c}^{c} h_i^{L} x_3^m dx_3 = \sum_{n=0}^{\infty} B^{(mn)}(\bar{\chi}_{ik} m_k^{(n)} + \bar{g}_{ikm} u_{km}^{(n)}), \qquad (8.2.4)$$

$$a_{ij}^{(m)} = \int_{-c}^{c} a_{ij} x_3^m dx_3 = \sum_{n=0}^{\infty} B^{(mn)} \bar{\beta}_{ijkl} m_{lk}^{(n)},$$

where we denoted

$$B^{(mn)} = B^{(nm)} = \int_{-c}^{c} x_3^m x_3^n dx_3$$

$$= \begin{cases} 2c^{m+n+1}/(m+n+1), & m+n \quad \text{even,} \\ 0, & m+n \quad \text{odd.} \end{cases} \qquad (8.2.5)$$

Equation (8.2.4) are essentially the two-dimensional plate constitutive relations. Then, from Eqs. (8.1.5)–(8.1.7) and (8.1.9), we obtain

$$f_j^{M(m)} = \int_{-c}^{c} M_i^0 h_{j,i}^M x_3^m dx_3$$

$$= \sum_{n=0}^{\infty} B^{(mn)} M_i^0 [h_{j,i}^{M(n)} + (n+1)\delta_{3i} h_j^{M(n+1)}], \qquad (8.2.6)$$

$$\tau_{ij}^{A(m)} = \int_{-c}^{c} \tau_{ij}^{A} x_3^m dx_3 = \frac{1}{2}(M_i^0 h_j^{L(m)} - M_j^0 h_i^{L(m)}), \qquad (8.2.7)$$

$$\tau_{ij}^{(m)} = \tau_{ij}^{S(m)} + \tau_{ij}^{A(m)}$$

$$= \sum_{n=0}^{\infty} B^{(mn)}(\bar{c}_{ijkm} u_{km}^{(n)} + \bar{e}_{kij} m_k^{(n)}) + \frac{1}{2}(M_i^0 h_j^{L(m)} - M_j^0 h_i^{L(m)}), \qquad (8.2.8)$$

$$b_i^{M(m)} = \int_{-c}^{c} b_i^M x_3^m dx_3$$

$$= \sum_{n=0}^{\infty} B^{(mn)}(h_i^{M(n)} + 4\pi m_i^{(m)} - 4\pi M_i^0 u_{jj}^{(n)}). \qquad (8.2.9)$$

The plate field equations are obtained by multiplying Eqs. (8.1.1), (8.1.3) and (8.1.4) by x_3^m and integrating them through the plate thickness. With integration by parts about x_3, this results in

$$\tau_{aj,a}^{(m)} - m\tau_{3j}^{(m-1)} + f_j^{(m)} + f_j^{M(m)} = \rho \sum_{n=0}^{\infty} B^{(mn)} \ddot{u}_j^{(n)}, \qquad (8.2.10)$$

$$b_{a,a}^{M(m)} - mb_3^{M(m-1)} + b^{M(m)} = 0, \qquad (8.2.11)$$

$$\varepsilon_{ijk} M_j^0 \left(\sum_{n=0}^{\infty} B^{(mn)} h_k^{M(n)} - a_{bk,b}^{(m)} + ma_{3k}^{(m-1)} - a_k^{(m)} + h_k^{L(m)} \right)$$

$$+ \varepsilon_{ijk} \sum_{n=0}^{\infty} B^{(mn)} m_j^{(n)} H_k^0 = \frac{1}{\gamma} \sum_{n=0}^{\infty} B^{(mn)} \dot{m}_i^{(n)}, \qquad (8.2.12)$$

where we have introduced a two-dimensional summation convention that subscripts a and b assume 1 and 2 only without 3. In Eqs. (8.2.10)–(8.2.12), the two-dimensional mechanical and magnetic loads are denoted by

$$f_j^{(m)} = [x_3^m \tau_{3j}]_{-c}^c + \int_{-c}^c f_j x_3^m dx_3,$$

$$b^{M(m)} = [x_3^m b_3^M]_{-c}^c, \qquad (8.2.13)$$

$$a_k^{(m)} = [x_3^m a_{3k}]_{-c}^c.$$

With successive substitutions, Eqs. (8.2.10)–(8.2.12) may be written as two-dimensional equations for $u_i^{(n)}$, $m_i^{(n)}$ and $\psi^{(n)}$.

8.3 Zero- and First-Order Equations

For a first-order plate theory we make the following truncations of the displacement, magnetization and magnetic potential:

$$u_k \cong u_k^{(0)} + x_3 u_k^{(1)} + x_3^2 \delta_{3k} u_3^{(2)},$$

$$m_k \cong m_k^{(0)} + x_3 m_k^{(1)}, \qquad (8.3.1)$$

$$\psi \cong \psi^{(0)} + x_3 \psi^{(1)},$$

where $u_1^{(0)}$ and $u_2^{(0)}$ are for extension, $u_3^{(0)}$ is for flexure, $u_1^{(1)}$ and $u_2^{(1)}$ are for the shear deformations accompanying flexure. $u_3^{(1)}$ and $u_3^{(2)}$ describe Poisson's effects in extension and flexure. $u_3^{(1)}$ and $u_3^{(2)}$ cannot be simply set to zero but will be eliminated later through the so-called stress relaxations for thin plates. Corresponding to Eq. (8.3.1), the gradient fields are

$$u_{km} \cong u_{km}^{(0)} + x_3 u_{km}^{(1)},$$

$$m_{km} \cong m_{km}^{(0)} + x_3 m_{km}^{(1)}, \qquad (8.3.2)$$

$$h_k^M \cong h_k^{M(0)} + x_3 h_k^{M(1)},$$

where

$$
\begin{aligned}
u_{11}^{(0)} &= u_{1,1}^{(0)}, & u_{12}^{(0)} &= u_{1,2}^{(0)}, & u_{13}^{(0)} &= u_1^{(1)}, \\
u_{21}^{(0)} &= u_{2,1}^{(0)}, & u_{22}^{(0)} &= u_{2,2}^{(0)}, & u_{23}^{(0)} &= u_2^{(1)}, \\
u_{31}^{(0)} &= u_{3,1}^{(0)}, & u_{32}^{(0)} &= u_{3,2}^{(0)}, & u_{33}^{(0)} &= u_3^{(1)},
\end{aligned}
\qquad (8.3.3)
$$

$$
\begin{aligned}
u_{11}^{(1)} &= u_{1,1}^{(1)}, & u_{12}^{(1)} &= u_{1,2}^{(1)}, & u_{13}^{(1)} &= 2u_1^{(2)} \cong 0, \\
u_{21}^{(1)} &= u_{2,1}^{(1)}, & u_{22}^{(1)} &= u_{2,2}^{(1)}, & u_{23}^{(1)} &= 2u_2^{(2)} \cong 0, \\
u_{31}^{(1)} &= u_{3,1}^{(1)}, & u_{32}^{(1)} &= u_{3,2}^{(1)}, & u_{33}^{(1)} &= 2u_3^{(2)},
\end{aligned}
\qquad (8.3.4)
$$

$$
\begin{aligned}
m_{11}^{(0)} &= m_{1,1}^{(0)}, & m_{12}^{(0)} &= m_{1,2}^{(0)}, & m_{13}^{(0)} &= m_1^{(1)}, \\
m_{21}^{(0)} &= m_{2,1}^{(0)}, & m_{22}^{(0)} &= m_{2,2}^{(0)}, & m_{23}^{(0)} &= m_2^{(1)}, \\
m_{31}^{(0)} &= m_{3,1}^{(0)}, & m_{32}^{(0)} &= m_{3,2}^{(0)}, & m_{33}^{(0)} &= m_3^{(1)},
\end{aligned}
\qquad (8.3.5)
$$

$$
\begin{aligned}
m_{11}^{(1)} &= m_{1,1}^{(1)}, & m_{12}^{(1)} &= m_{1,2}^{(1)}, & m_{13}^{(1)} &= 2m_1^{(2)} \cong 0, \\
m_{21}^{(1)} &= m_{2,1}^{(1)}, & m_{22}^{(1)} &= m_{2,2}^{(1)}, & m_{23}^{(1)} &= 2m_2^{(2)} \cong 0, \\
m_{31}^{(1)} &= m_{3,1}^{(1)}, & m_{32}^{(1)} &= m_{3,2}^{(1)}, & m_{33}^{(1)} &= 2m_3^{(2)} \cong 0,
\end{aligned}
\qquad (8.3.6)
$$

$$
\begin{aligned}
h_1^{M(0)} &= -\psi_{,1}^{(0)}, & h_2^{M(0)} &= -\psi_{,2}^{(0)}, & h_3^{M(0)} &= -\psi^{(1)}, \\
h_1^{M(1)} &= -\psi_{,1}^{(1)}, & h_2^{M(1)} &= -\psi_{,2}^{(1)}, & h_3^{M(1)} &= -2\psi^{(2)} \cong 0.
\end{aligned}
\qquad (8.3.7)
$$

The zero- and first-order plate constitutive relations are

$$
\begin{aligned}
\tau_{ij}^{S(0)} &= 2c(\bar{c}_{ijkm} u_{km}^{(0)} + \bar{h}_{kij} m_k^{(0)}), \\
h_i^{L(0)} &= 2c(\bar{\chi}_{ik} m_k^{(0)} + \bar{g}_{ikm} u_{km}^{(0)}), \\
a_{ij}^{(0)} &= 2c\bar{\beta}_{ijkl} m_{lk}^{(0)},
\end{aligned}
\qquad (8.3.8)
$$

$$
\begin{aligned}
\tau_{ij}^{S(1)} &= I(\bar{c}_{ijkm} u_{km}^{(1)} + \bar{h}_{kij} m_k^{(1)}), \\
h_i^{L(1)} &= I(\bar{\chi}_{ik} m_k^{(1)} + \bar{g}_{ikm} u_{km}^{(1)}), \\
a_{ij}^{(1)} &= I\bar{\beta}_{ijkl} m_{lk}^{(1)}, \quad I = \frac{2c^3}{3},
\end{aligned}
\qquad (8.3.9)
$$

$$\tau_{ij}^{A(0)} = \frac{1}{2}(M_i^0 h_j^{L(0)} - M_j^0 h_i^{L(0)}),$$

$$\tau_{ij}^{A(1)} = \frac{1}{2}(M_i^0 h_j^{L(1)} - M_j^0 h_i^{L(1)}),$$

(8.3.10)

$$\tau_{ij}^{(0)} = 2c(\bar{c}_{ijkm} u_{km}^{(0)} + \bar{h}_{kij} m_k^{(0)}) + \frac{1}{2}(M_i^0 h_j^{L(0)} - M_j^0 h_i^{L(0)}),$$

$$\tau_{ij}^{(1)} = I(\bar{c}_{ijkm} u_{km}^{(1)} + \bar{h}_{kij} m_k^{(1)}) + \frac{1}{2}(M_i^0 h_j^{L(1)} - M_j^0 h_i^{L(1)}),$$

(8.3.11)

$$b_i^{M(0)} = 2c(h_i^{M(0)} + 4\pi m_i^{(0)} - 4\pi M_i^0 u_{jj}^{(0)}),$$

$$b_i^{M(1)} = I(h_i^{M(1)} + 4\pi m_i^{(1)} - 4\pi M_i^0 u_{jj}^{(1)}).$$

(8.3.12)

For the extension and bending of thin plates, the following stress components are relatively small and are set to zero [4] (stress relaxation):

$$\tau_{33}^{(0)} = 0,$$

(8.3.13)

$$\tau_{31}^{(1)} = \tau_{32}^{(1)} = \tau_{33}^{(1)} = 0.$$

(8.3.14)

Equation (8.3.13) is used to eliminate $u_{33}^{(0)} = u_3^{(1)}$ which may appear in Eqs. (8.3.11)$_1$ and (8.3.12)$_1$. Equation (8.3.14) is used to eliminate $u_{31}^{(1)} = u_{3,1}^{(1)}$, $u_{32}^{(1)} = u_{3,2}^{(1)}$ and $u_{33}^{(1)} = 2u_3^{(2)}$ which may be present in Eqs. (8.3.11)$_2$ and (8.3.12)$_2$. These will be carried out in Section 8.4 for cubic crystals specifically. The zero- and first-order field equations are

$$\tau_{aj,a}^{(0)} + f_j^{(0)} + f_j^{M(0)} = \rho 2c\ddot{u}_j^{(0)},$$

$$\tau_{ab,a}^{(1)} - \tau_{3b}^{(0)} + f_b^{(1)} + f_b^{M(1)} = \rho I\ddot{u}_b^{(1)},$$

(8.3.15)

$$b_{a,a}^{M(0)} + b^{M(0)} = 0,$$

$$b_{a,a}^{M(1)} - b_3^{M(0)} + b^{M(1)} = 0,$$

(8.3.16)

$$\varepsilon_{ijk} M_j^0 (2ch_k^{M(0)} - a_{bk,b}^{(0)} - a_k^{(0)} + h_k^{L(0)})$$

$$+ \varepsilon_{ijk} 2cm_j^{(0)} H_k^0 = \frac{1}{\gamma} 2c\dot{m}_i^{(0)},$$

$$\varepsilon_{ijk} M_j^0 (Ih_k^{M(1)} - a_{bk,b}^{(1)} + a_{3k}^{(0)} - a_k^{(1)} + h_k^{L(1)})$$

$$+ \varepsilon_{ijk} Im_j^{(1)} H_k^0 = \frac{1}{\gamma} I\dot{m}_i^{(1)}.$$

(8.3.17)

Equations (8.3.15)–(8.3.17) may be written as thirteen equations for the thirteen components of $u_i^{(0)}$, $u_a^{(1)}$, $m_i^{(0)}$, $m_i^{(1)}$, $\psi^{(0)}$ and $\psi^{(1)}$.

8.4 Cubic Crystals with Thickness Magnetization

Consider a plate of YIG which is a cubic crystal of class (m3m). Let the spontaneous magnetization $\mathbf{M}^0$ (and $\mathbf{H}^0$) be along the x_3 axis. $\mathbf{M} = \mathbf{M}^0 + \mathbf{m}$ and $\mathbf{m}$ is small. In this case $m_3 = 0$ because of the saturation condition $\mathbf{M} \cdot \mathbf{M} = (M^0)^2$ which implies that $\mathbf{M}^0 \cdot \mathbf{m} = 0$. In addition, τ^A is found to be relatively small and is negligible compared to τ^S [18]. The three-dimensional constitutive relations below are from Section 6.2:

$$
\begin{aligned}
\tau_1 = \tau_{11} &= c_{11}u_{1,1} + c_{12}u_{2,2} + c_{12}u_{3,3}, \\
\tau_2 = \tau_{22} &= c_{12}u_{1,1} + c_{11}u_{2,2} + c_{12}u_{3,3}, \\
\tau_3 = \tau_{33} &= c_{12}u_{1,1} + c_{12}u_{2,2} + c_{11}u_{3,3}, \\
\tau_4 = \tau_{23}^S &= c_{44}(u_{2,3} + u_{3,2}) + 2b_{44}M^0 m_2, \\
\tau_5 = \tau_{31}^S &= c_{44}(u_{1,3} + u_{3,1}) + 2b_{44}M^0 m_1, \\
\tau_6 = \tau_{12}^S &= c_{44}(u_{1,2} + u_{2,1}),
\end{aligned} \tag{8.4.1}
$$

$$
\begin{aligned}
h_1^L &= -\chi(M^0)^2 m_1 - 2b_{44}M^0(u_{1,3} + u_{3,1}), \\
h_2^L &= -\chi(M^0)^2 m_2 - 2b_{44}M^0(u_{2,3} + u_{3,2}), \\
h_3^L &= 0,
\end{aligned} \tag{8.4.2}
$$

$$
a_{ib} = -2\alpha_{11}m_{b,i}. \tag{8.4.3}
$$

The corresponding two-dimensional constitutive equations for the zero-order and first-order plate equations, after the stress relaxations in Eqs. (8.3.13) and (8.3.14), are found to be

$$
\begin{aligned}
\tau_1^{(0)} &= 2c(c_{11}' u_{11}^{(0)} + c_{12}' u_{22}^{(0)}), \\
\tau_2^{(0)} &= 2c(c_{12}' u_{11}^{(0)} + c_{11}' u_{22}^{(0)}), \\
\tau_4^{(0)} &= 2c[\kappa^2 c_{44}(u_{23}^{(0)} + u_{32}^{(0)}) + 2b_{44}M^0 m_2^{(0)}], \\
\tau_5^{(0)} &= 2c[\kappa^2 c_{44}(u_{13}^{(0)} + u_{31}^{(0)}) + 2b_{44}M^0 m_1^{(0)}], \\
\tau_6^{(0)} &= 2cc_{44}(u_{12}^{(0)} + u_{21}^{(0)}),
\end{aligned} \tag{8.4.4}
$$

$$\tau_1^{(1)} = I(c_{11}' u_{11}^{(1)} + c_{12}' u_{22}^{(1)}),$$
$$\tau_2^{(1)} = I(c_{12}' u_{11}^{(1)} + c_{11}' u_{22}^{(1)}), \tag{8.4.5}$$
$$\tau_6^{(1)} = Ic_{44}(u_{12}^{(1)} + u_{21}^{(1)}),$$

$$b_a^{M(0)} = 2c(h_a^{M(0)} + 4\pi m_a^{(0)}),$$
$$b_3^{M(0)} = 2c(h_3^{M(0)} - 4\pi M^0 u_{jj}^{(0)}), \tag{8.4.6}$$
$$b_a^{M(1)} = I(h_a^{M(1)} + 4\pi m_a^{(1)}),$$

$$h_1^{L(0)} = -2c[\chi(M^0)^2 m_1^{(0)} + 2b_{44} M^0 (u_{13}^{(0)} + u_{31}^{(0)})],$$
$$h_2^{L(0)} = -2c[\chi(M^0)^2 m_2^{(0)} + 2b_{44} M^0 (u_{23}^{(0)} + u_{32}^{(0)})], \tag{8.4.7}$$

$$h_1^{L(1)} = -I(\chi - 4b_{44}^2/c_{44})(M^0)^2 m_1^{(1)},$$
$$h_2^{L(1)} = -I(\chi - 4b_{44}^2/c_{44})(M^0)^2 m_2^{(1)}, \tag{8.4.8}$$

$$a_{ib}^{(0)} = -4c\alpha_{11} m_{bi}^{(0)}, \quad a_{ab}^{(1)} = -2I\alpha_{11} m_{ba}^{(1)}, \tag{8.4.9}$$

$$f_1^{M(0)} = 2cM_3^0 h_1^{M(1)},$$
$$f_2^{M(0)} = 2cM_3^0 h_2^{M(1)}, \tag{8.4.10}$$

where the effective two-dimensional elastic constants for thin plates after the stress relaxations are denoted by

$$c_{11}' = c_{11} - \frac{c_{12}^2}{c_{11}}, \quad c_{12}' = c_{12} - \frac{c_{12}^2}{c_{11}}. \tag{8.4.11}$$

In Eqs. $(8.4.4)_{3,4}$, we have inserted a shear correction factor κ. An approximate value of κ may be taken as $\kappa^2 \cong \pi^2/12$ [4, 38–41]. Shear correction factors may be introduced in ways more sophisticated [4, 38–41] than what is in Eqs. $(8.4.4)_{3,4}$. Substituting the above constitutive relations into the field equations in Eqs. (8.3.15)–(8.3.17), we obtain two uncoupled sets of equations. One consists of six equations for bending with shear deformations in terms of $u_3^{(0)}$, $u_1^{(1)}$ and $u_2^{(1)}$ which are coupled to $\psi^{(0)}$, $m_1^{(0)}$ and $m_2^{(0)}$:

$$\kappa^2 c_{44}(u_{3,aa}^{(0)} + u_{a,a}^{(1)}) + 2b_{44} M^0 m_{a,a}^{(0)} + f_3^{(0)}/2c = \rho \ddot{u}_3^{(0)}, \tag{8.4.12}$$

$$c'_{11}u^{(1)}_{1,11} + c_{44}u^{(1)}_{1,22} + (c'_{12} + c_{44})u^{(1)}_{2,12}$$

$$- \frac{3}{c^2}[\kappa^2 c_{44}(u^{(1)}_1 + u^{(0)}_{3,1}) + 2b_{44}M^0 m^{(0)}_1] + f^{(1)}_1/I = \rho\ddot{u}^{(1)}_1,$$

$$(c'_{12} + c_{44})u^{(1)}_{1,12} + c_{44}u^{(1)}_{2,11} + c'_{11}u^{(1)}_{2,22}$$

$$- \frac{3}{c^2}[\kappa^2 c_{44}(u^{(1)}_2 + u^{(0)}_{3,2}) + 2b_{44}M^0 m^{(0)}_2] + f^{(1)}_2/I = \rho\ddot{u}^{(1)}_2, \tag{8.4.13}$$

$$2c(-\psi^{(0)}_{,aa} + 4\pi m^{(0)}_{a,a}) + b^{M(0)} = 0, \tag{8.4.14}$$

$$-M^0[-\psi^{(0)}_{,2} + 2\alpha_{11}m^{(0)}_{2,aa} - \chi(M^0)^2 m^{(0)}_2 - 2b_{44}M^0(u^{(1)}_2 + u^{(0)}_{3,2})]$$

$$+ M^0 a^{(0)}_2/2c + m^{(0)}_2 H^0 = \frac{1}{\gamma}\dot{m}^{(0)}_1,$$

$$M^0[-\psi^{(0)}_{,1} + 2\alpha_{11}m^{(0)}_{1,aa} - \chi(M^0)^2 m^{(0)}_1 - 2b_{44}M^0(u^{(1)}_1 + u^{(0)}_{3,1})]$$

$$- M^0 a^{(0)}_1/2c - m^{(0)}_1 H^0 = \frac{1}{\gamma}\dot{m}^{(0)}_2. \tag{8.4.15}$$

The other has five equations for extension in terms of $u^{(0)}_1$ and $u^{(0)}_2$ which are coupled to $\psi^{(1)}$, $m^{(1)}_1$ and $m^{(1)}_2$:

$$c'_{11}u^{(0)}_{1,11} + c_{44}u^{(0)}_{1,22} + (c'_{12} + c_{44})u^{(0)}_{2,12} - M^0\psi^{(1)}_{,1} + f^{(0)}_1/2c = \rho\ddot{u}^{(0)}_1,$$

$$(c'_{12} + c_{44})u^{(0)}_{1,12} + c_{44}u^{(0)}_{2,11} + c'_{11}u^{(0)}_{2,22} - M^0\psi^{(1)}_{,2} + f^{(0)}_2/2c = \rho\ddot{u}^{(0)}_2,$$

$$\tag{8.4.16}$$

$$I(-\psi^{(1)}_{,aa} + 4\pi m^{(1)}_{a,a}) - 2c\left(-\psi^{(1)} - 4\pi M^0\frac{c_{11} - c_{12}}{c_{11}}u^{(0)}_{a,a}\right) + b^{M(1)} = 0,$$

$$\tag{8.4.17}$$

$$-M^0\left[-\psi^{(1)}_{,2} + 2\alpha_{11}m^{(1)}_{2,aa} - \frac{6}{c^2}\alpha_{11}m^{(1)}_2 - (\chi - 4b^2_{44}/c_{44})(M^0)^2 m^{(1)}_2\right]$$

$$+ M^0 a^{(1)}_2/I + m^{(1)}_2 H^0 = \frac{1}{\gamma}\dot{m}^{(1)}_1,$$

$$M^0\left[-\psi^{(1)}_{,1} + 2\alpha_{11}m^{(1)}_{1,aa} - \frac{6}{c^2}\alpha_{11}m^{(1)}_1 - (\chi - 4b^2_{44}/c_{44})(M^0)^2 m^{(1)}_1\right] \tag{8.4.18}$$

$$- M^0 a^{(1)}_1/I - m^{(1)}_1 H^0 = \frac{1}{\gamma}\dot{m}^{(1)}_2.$$

8.5   Thickness Vibrations

Thickness vibrations of plates are independent of the in-plane coordinates x_1 and x_2. Consider the set of equations in Eqs. (8.4.12)–(8.4.15) first. The mechanical and magnetic loads are assumed to be zero. Equations (8.4.12) and (8.4.14) become trivial. Equations (8.4.13) and (8.4.15) reduce to the following four equations:

$$-3(\kappa^2 c_{44} u_1^{(1)} + 2b_{44} M^0 m_1^{(0)}) = \rho c^2 \ddot{u}_1^{(1)},$$
$$-3(\kappa^2 c_{44} u_2^{(1)} + 2b_{44} M^0 m_2^{(0)}) = \rho c^2 \ddot{u}_2^{(1)}, \tag{8.5.1}$$

$$\gamma M^0 (P m_2^{(0)} + 2b_{44} M^0 u_2^{(1)}) = \dot{m}_1^{(0)},$$
$$-\gamma M^0 (P m_1^{(0)} + 2b_{44} M^0 u_1^{(1)}) = \dot{m}_2^{(0)}, \tag{8.5.2}$$

where

$$P = \frac{H^0}{M^0} + K, \quad K = \chi (M^0)^2. \tag{8.5.3}$$

Let

$$u_1^{(1)} = A_1 \sin \omega t, \quad u_2^{(1)} = A_2 \cos \omega t,$$
$$m_1^{(0)} = C_1 \sin \omega t, \quad m_2^{(0)} = C_2 \cos \omega t. \tag{8.5.4}$$

Substituting Eq. (8.5.4) into Eqs. (8.5.1) and (8.5.2), we obtain

$$\rho c^2 \omega^2 A_1 - 3(\kappa^2 c_{44} A_1 + 2b_{44} M^0 C_1) = 0,$$
$$\rho c^2 \omega^2 A_2 - 3(\kappa^2 c_{44} A_2 + 2b_{44} M^0 C_2) = 0, \tag{8.5.5}$$

$$\gamma M^0 (P C_2 + 2b_{44} M^0 A_2) - \omega C_1 = 0,$$
$$\gamma M^0 (P C_1 + 2b_{44} M^0 A_1) - \omega C_2 = 0. \tag{8.5.6}$$

For nontrivial solutions,

$$\begin{vmatrix} \rho c^2 \omega^2 - 3\kappa^2 c_{44} & 0 & -6b_{44}M^0 & 0 \\ 0 & \rho c^2 \omega^2 - 3\kappa^2 c_{44} & 0 & -6b_{44}M^0 \\ 0 & 2b_{44}\gamma(M^0)^2 & -\omega & \gamma M^0 P \\ 2b_{44}\gamma(M^0)^2 & 0 & \gamma M^0 P & -\omega \end{vmatrix} = 0. \tag{8.5.7}$$

In the special case, when $b_{44} = 0$, the magnetoelastic coupling disappears and Eq. (8.5.7) leads to the following cutoff frequencies for uncoupled elastic waves:

$$\omega_c^2 = \frac{3\kappa^2 c_{44}}{\rho c^2} = \frac{\pi^2}{12}\frac{3c_{44}}{\rho c^2} = \frac{\pi^2}{4c^2}\frac{c_{44}}{\rho}, \tag{8.5.8}$$

which agrees with Eq. (1.8.5) when $n = 1$, and uncoupled spin waves with

$$\omega_c = |\gamma M^0 P|, \tag{8.5.9}$$

which agrees with Eq. (6.3.19).

Next consider the other set of plate equations in Eqs. (8.4.16)–(8.4.18). For thickness vibrations, Eqs. (8.4.16) and (8.4.17) become trivial. Equation (8.4.18) reduces to

$$\frac{3}{c^2}\alpha m_2^{(1)} - \frac{e^2}{c_{44}}m_2^{(1)} + m_2^{(1)}P = \frac{1}{\gamma M^0}\dot{m}_1^{(1)},$$
$$-\frac{3}{c^2}\alpha m_1^{(1)} + \frac{e^2}{c_{44}}em_1^{(1)} - m_1^{(1)}P = \frac{1}{\gamma M^0}\dot{m}_2^{(1)}, \tag{8.5.10}$$

where

$$e = 2b_{44}M^0, \quad \alpha = 2\alpha_{11},$$
$$P = H^0/M^0 + K, \quad K = \chi(M^0)^2. \tag{8.5.11}$$

Let

$$m_1^{(1)} = C_1 \sin\omega t, \quad m_2^{(1)} = C_2 \cos\omega t. \tag{8.5.12}$$

Substituting Eq. (8.5.12) into Eq. (8.5.10), we obtain

$$\frac{3}{c^2}\alpha C_2 - \frac{e^2}{c_{44}}C_2 + PC_2 = \frac{1}{\gamma M^0}\omega C_1,$$
$$-\frac{3}{c^2}\alpha C_1 + \frac{e^2}{c_{44}}C_1 - PC_1 = -\frac{1}{\gamma M^0}\omega C_2. \tag{8.5.13}$$

For nontrivial solutions,

$$\begin{vmatrix} -\dfrac{1}{\gamma M^0}\omega & \dfrac{3}{c^2}\alpha - \dfrac{e^2}{c_{44}} + P \\[3mm] \dfrac{3}{c^2}\alpha - \dfrac{e^2}{c_{44}} + P & -\dfrac{1}{\gamma M^0}\omega \end{vmatrix} = 0, \tag{8.5.14}$$

or

$$\frac{1}{\gamma M^0}\omega = \pm\left(\frac{3}{c^2}\alpha - \frac{e^2}{c_{44}} + P\right). \tag{8.5.15}$$

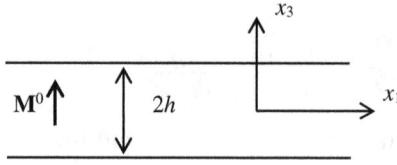

Fig. 8.2. A plate and coordinate system.

8.6 Waves in Plates

In this section, we analyze waves in the same unbounded plate (see Fig. 8.2) as in Section 6.4 under the same approximation of $\psi \cong 0$ using the two-dimensional plate equations in Eqs. (8.4.12)–(8.4.18). We study the so-called straight-crested waves propagating in the x_1 direction without x_2 dependence. The waves are labeled by

$$u_1^{(0)} = \text{extension (Ext)},$$
$$u_2^{(0)} = \text{face shear (FS)}, \qquad (8.6.1)$$
$$u_3^{(0)} = \text{flexure (F)},$$

$$u_1^{(1)} = \text{thickness shear (TSh)},$$
$$u_2^{(1)} = \text{thickness twist (TT)}, \qquad (8.6.2)$$

$$m_1^{(0)}, m_2^{(0)} = \text{zero-order spin (Spin-0)},$$
$$m_1^{(1)}, m_2^{(2)} = \text{first-order spin (Spin-1)}. \qquad (8.6.3)$$

The displacements and deformations associated with the elastic waves described by Eqs. (8.6.1) and (8.6.2) are shown in Fig. 8.3.

Let the dimensionless frequency and wave number be

$$\Omega = \omega \left/ \left(\frac{\pi}{2c} \sqrt{\frac{c_{44}}{\rho}} \right) \right., \quad X = \xi \left/ \left(\frac{\pi}{2c} \right) \right.. \qquad (8.6.4)$$

Figure 8.4 shows the dispersion curves obtained from the plate equations [42]. (a) and (b) are for the same group of waves but (a) does not have magnetoelastic coupling ($b_{44} = 0$) while (b) has ($b_{44} \neq 0$). (c) is for the other group of waves. The main difference between the results from the plate equations and those from the three-dimensional equations in Figs. 6.4 and 6.5 is that the plate equations describe the zero- and first-order fields only and do not have the dispersion curves of higher-order waves.

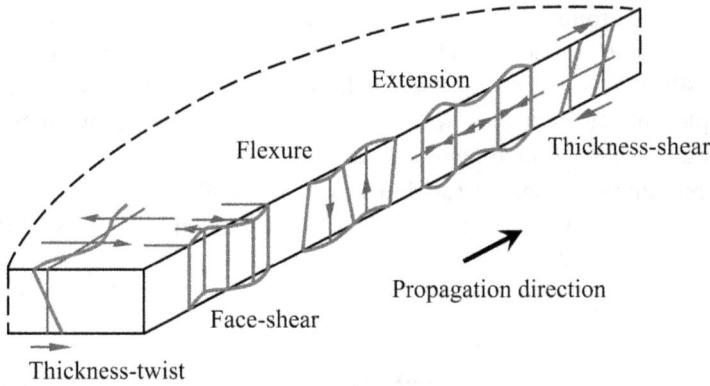

Fig. 8.3. Zero- and first-order elastic waves propagating in the x_1 direction.

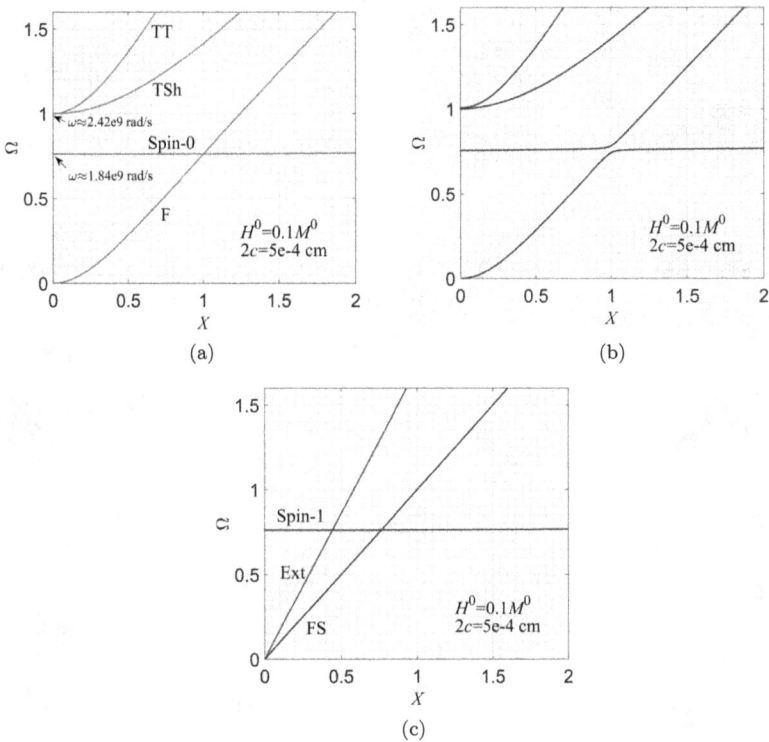

Fig. 8.4. Dispersion curves from plate equations. (a) TS, TT, F and Spin-0 waves when $b_{44} = 0$. (b) TS, TT, F and Spin-0 when $b_{44} \neq 0$. (c) Ext, FS and Spin-1 waves when $b_{44} \neq 0$.

Since the group of TS, TT, F and Spin-0 waves shows magnetoelastic coupling, we examine the effects of some parameters on these waves. Figure 8.5 differs from Fig. 8.4 by the plate thickness $2c$ [42]. (a) is for uncoupled elastic and spin waves ($b_{44} = 0$). (b) is with magnetoelastic coupling ($b_{44} \neq 0$). The dimensionless cutoff frequencies of the TS and TT waves remain the same, but the dimensionless cutoff frequency of the Spin-0 wave increases. Since the dimensionless cutoff frequency of the Spin-0 wave

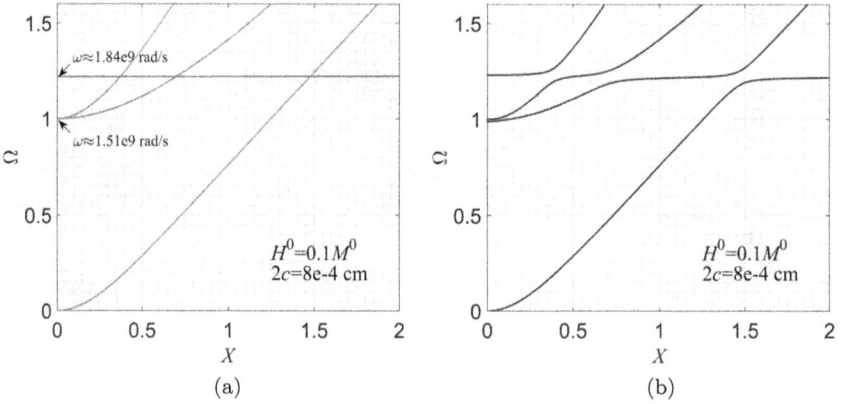

Fig. 8.5. Dispersion curves of TS, TT, F and Spin-0 waves. (a) $H^0 = 0.1M^0$, $2c = 8 \times 10^{-4}$ cm and $b_{44} = 0$. (b) $H^0 = 0.1M^0$, $2c = 8 \times 10^{-4}$ cm and $b_{44} \neq 0$.

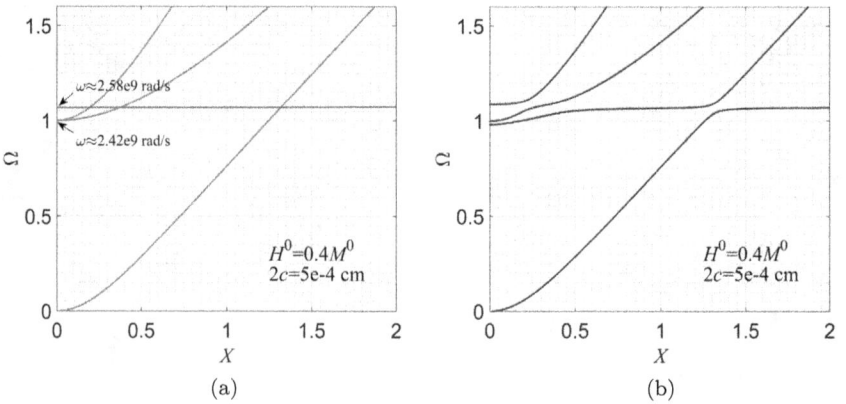

Fig. 8.6. Dispersion curves of TS, TT, F and Spin-0 waves. (a) $H^0 = 0.4M^0$, $2c = 5 \times 10^{-4}$ cm and $b_{44} = 0$. (b) $H^0 = 0.4M^0$, $2c = 5 \times 10^{-4}$ cm and $b_{44} \neq 0$.

is now higher than the dimensionless cutoff frequencies of the TS and TT waves, more magnetoelastic couplings exist among these waves.

Figure 8.6 differs from Fig. 8.4 by the biasing magnetic field H_0 [42]. As H_0 increases, the dimensionless cutoff frequencies of the TS and TT waves change very little, but that of the Spin-0 wave increases significantly to cause more magnetoelastic couplings.

Chapter 9

ELASTIC BEAMS WITH POINT MAGNETS

Elastic beams with discrete or point magnets at their ends or in the interior are common structures in applications [43–45]. This chapter establishes one-dimensional models for elastic beams with point magnets in extension, bending and torsion which may be coupled by the point magnets when an external magnetic field is applied. The point magnets are assumed to be small and rigid.

9.1 Equations of a Small and Rigid Magnet

Consider a small and rigid ferromagnet with a volume V and mass density ρ. Based on the two-continuum model [16], it is assumed that the magnet consists of a rigid lattice continuum and a massless spin continuum. We denote the total mass of the magnet by m and consider the magnetization vector $\mathbf{M}$ to be its average over V for a small magnet:

$$m = \int_V \rho(\mathbf{x})dV,$$

$$\mathbf{M}(t) = \tfrac{1}{V}\int_V \mathbf{M}(\mathbf{x}, t)dV.$$

$$(9.1.1)$$

The centroidal velocity of the magnet is denoted by $\mathbf{v}_G$ with respect to an inertial reference frame. We also introduce a centroidal reference frame for the magnet. The centroidal frame is in translation only without rotation. The rotations of the lattice continuum and the magnetic moment vector of the spin continuum with respect to the centroidal system may be different, and are both assumed to be small. They are described by small angular displacements $\boldsymbol{\delta\theta}^l$ and $\boldsymbol{\delta\theta}^s$, respectively. The angular rates of the rotations of the lattice and the magnetic moment vector of the spin are denoted by

$$\boldsymbol{\omega}^l = \frac{\boldsymbol{\delta\theta}^l}{\delta t}, \quad \boldsymbol{\omega}^s = \frac{\boldsymbol{\delta\theta}^s}{\delta t}. \tag{9.1.2}$$

The mass moment of inertia of the lattice with respect to its translational centroidal reference frame is denoted by a tensor $\mathbf{I}_G$. In general, $\mathbf{I}_G$ may be time dependent when the magnet is rotating with respect to its centroidal frame. We assume a spherical magnet whose $(\mathbf{I}_G)_{ij} = I_G\delta_{ij}$ or $\mathbf{I}_G = I_G\mathbf{1}$, where $\mathbf{1}$ is the second-order unit tensor and I_G is a constant. Then the angular momentum of the lattice is given by $\mathbf{I}_G \cdot \boldsymbol{\omega}^l = I_G\boldsymbol{\omega}^l$. The linear momentum equation for the magnet is

$$\frac{d}{dt}(m\mathbf{v}_G) = \Sigma\mathbf{F}, \tag{9.1.3}$$

where $\Sigma\mathbf{F}$ is the total force acting on the magnet. The angular momentum equations for the lattice and the spin separately are as follows:

$$\frac{d}{dt}(I_G\boldsymbol{\omega}^l) = \boldsymbol{\Gamma}, \tag{9.1.4}$$

$$\frac{1}{\gamma}\frac{d\mathbf{M}}{dt} = \mathbf{M} \times \mathbf{B}^{e\!f\!f}, \tag{9.1.5}$$

where $\boldsymbol{\Gamma}$ is the total couple on the lattice.

The interaction between the lattice and the spin is described by a local magnetic induction $\mathbf{B}^L$. The couple on the spin produced by $\mathbf{B}^L$ is $\mathbf{M} \times \mathbf{B}^L$. When unperturbed, the spontaneous magnetization $\mathbf{M}$ is along a direction of the lattice called an easy axis. Under a small disturbance, $\mathbf{M}$ deviates from the easy axis a little and experiences a restoring couple toward the easy axis. Effectively, $\mathbf{M} \times \mathbf{B}^L$ plays the role of the restoring couple on the spin. For a simple description of the restoring couple we assume it is related to the relative rotation of the spin with respect to the lattice linearly, i.e.,

$$\mathbf{M} \times \mathbf{B}^L = -\lambda M_s^2(\boldsymbol{\delta\theta}^s - \boldsymbol{\delta\theta}^l) = -\lambda M_s^2 \boldsymbol{\delta\theta}^{s/l},$$
$$\boldsymbol{\delta\theta}^{s/l} = \boldsymbol{\delta\theta}^s - \boldsymbol{\delta\theta}^l, \tag{9.1.6}$$

where λ is a material parameter and M_s is the saturation magnetization. Multiplying both sides of Eq. (9.1.6) by $\mathbf{M}$ through a cross product, we have

$$(\mathbf{M} \times \mathbf{B}^L) \times \mathbf{M} = -\lambda M_s^2 \boldsymbol{\delta\theta}^{s/l} \times \mathbf{M}. \tag{9.1.7}$$

With the following vector identity [14]:

$$\mathbf{A} \times (\mathbf{B} \times \mathbf{C}) = (\mathbf{A} \cdot \mathbf{C})\mathbf{B} - (\mathbf{A} \cdot \mathbf{B})\mathbf{C}, \tag{9.1.8}$$

we write Eq. (9.1.7) as

$$\begin{aligned}
(\mathbf{M} \times \mathbf{B}^L) \times \mathbf{M} &= -\mathbf{M} \times (\mathbf{M} \times \mathbf{B}^L) = \mathbf{M} \times (\mathbf{B}^L \times \mathbf{M}) \\
&= (\mathbf{M} \cdot \mathbf{M})\mathbf{B}^L - (\mathbf{M} \cdot \mathbf{B}^L)\mathbf{M} \\
&= M_s^2 \mathbf{B}^L - 0\mathbf{M} = M_s^2 \mathbf{B}^L = -\lambda M_s^2 \boldsymbol{\delta\theta}^{s/l} \times \mathbf{M}, \quad (9.1.9)
\end{aligned}$$

where $\mathbf{M} \cdot \mathbf{B}^L = 0$ has been used (see Eq. (4.2.4)). Hence,

$$\mathbf{B}^L = -\lambda \boldsymbol{\delta\theta}^{s/l} \times \mathbf{M}, \tag{9.1.10}$$

which serves as the constitutive relation for $\mathbf{B}^L$. Equation (9.1.10) can be written as

$$\mathbf{B}^L = -\lambda(\boldsymbol{\delta}\mathbf{M})^{s/l}, \tag{9.1.11}$$

where

$$(\boldsymbol{\delta}\mathbf{M})^{s/l} = \boldsymbol{\delta\theta}^{s/l} \times \mathbf{M}. \tag{9.1.12}$$

When $\lambda = 0$, we have $\mathbf{B}^L = 0$. In this case, there is no restoring couple. Any axis is an easy axis. When $\lambda = \infty$, from Eq. (9.1.6)$_1$, we have $\boldsymbol{\delta\theta}^{s/l} = 0$. In this case, there is no relative rotation between the spin and the lattice or, in other words, the spin is fixed to the lattice. Using Eq. (9.1.11), we can write the restoring couple on $\mathbf{M}$ as

$$\mathbf{M} \times \mathbf{B}^L = -\lambda \mathbf{M} \times (\boldsymbol{\delta}\mathbf{M})^{s/l}. \tag{9.1.13}$$

9.2 One-Dimensional Equations for Elastic Beams

Consider the elastic beam in Fig. 9.1. It has a circular cross-section so that the one-dimensional theory for torsion in conventional mechanics of materials applies [46]. The spatial coordinates are x_k or (x, y, z). The x-axis goes through the center of the circular cross-section. The material is assumed to be isotropic. We derive one-dimensional theories for extension, bending in the (x, y) and (x, z) planes as well as torsion in the following.

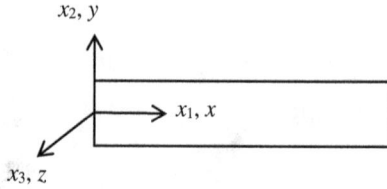

Fig. 9.1. An elastic beam with a circular cross-section.

For a one-dimensional model of the beam, the flexural displacements u_2 and u_3, axial displacement u_1 for extension and bending, and the angle of twist ψ for torsion are approximated by [46–48]

$$u_2|_{y=z=0} = v(x,t), \quad u_3|_{y=z=0} = w(x,t),$$

$$u_1(\mathbf{x},t) = u(x,t) - yv' - zw', \tag{9.2.1}$$

$$\psi = \psi(x,t).$$

The axial strain ε and the shear strain γ in torsion are related to u, v, w and ψ by

$$\varepsilon = u' - yv'' - zw'', \quad \gamma = r\psi', \tag{9.2.2}$$

where a prime represents a differentiation with respect to x, the axial coordinate. r and θ are polar coordinates within the circular cross-section. The axial stress σ and the shear stress τ in torsion can be expressed in terms of ε and γ through stress-strain relations as

$$\sigma = E\varepsilon = E(u' - yv'' - zw''),$$
$$\tau = G\gamma = Gr\psi', \tag{9.2.3}$$

where E is Young's modulus and G the shear modulus. The internal forces and moments over a typical cross-section of the beam are shown in Fig. 9.2. In Fig. 9.2, the moment vectors are represented by double arrows to distinguish them from force vectors. The axial force N, twisting moment or torque $\mathcal{M}_x$ for torsion, and bending moments $\mathcal{M}_y$ and $\mathcal{M}_z$ are obtained by integrating the relevant stresses or their products with y, z or r over a cross-section [46–48]. They have the following expressions:

$$N = EAu', \quad \mathcal{M}_x = GI_p\psi',$$
$$\mathcal{M}_y = -EIw'', \quad \mathcal{M}_z = EIv'', \tag{9.2.4}$$

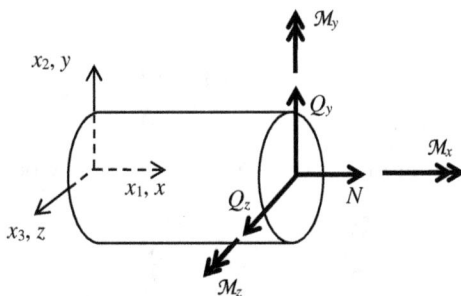

Fig. 9.2. Internal forces and moments over a cross-section.

where A is the area of the cross-section. I and I_p are the moment of inertia and polar moment of inertia of the cross-section. Let the radius of the circular cross-section be a. Then

$$A = \pi a^2, \quad I = \frac{\pi}{4} a^4, \quad I_p = \frac{\pi}{2} a^4. \tag{9.2.5}$$

The transverse shear forces Q_y and Q_z related to bending are determined from the following shear force-bending moment relations [46–48]:

$$Q_y = -\mathcal{M}'_z = -EIv''', \quad Q_z = \mathcal{M}'_y = -EIw'''. \tag{9.2.6}$$

The equations for extension, torsion, and deflections in the y and z directions are [46–48]:

$$N' + F_x = \rho A \ddot{u}, \quad \mathcal{M}'_x + m_x = \rho I_p \ddot{\psi},$$
$$Q'_y + F_y = \rho A \ddot{v}, \quad Q'_z + F_z = \rho A \ddot{w}, \tag{9.2.7}$$

where ρ is the mass density. **F** is distributed force per unit length of the beam. m_x is distributed twisting moment per unit length. With substitutions from Eqs. (9.2.6) and (9.2.4), we can write (9.2.7) as four uncoupled equations for u, ψ, v and w:

$$EAu'' + F_x = \rho A \ddot{u}, \tag{9.2.8}$$

$$GI_p\psi'' + m_x = \rho I_p \ddot{\psi}, \tag{9.2.9}$$

$$-EIv'''' + F_y = \rho A \ddot{v}, \tag{9.2.10}$$

$$-EIw'''' + F_z = \rho A \ddot{w}, \tag{9.2.11}$$

where a homogeneous beam with uniform geometric and material properties has been assumed. The small rotations of a cross-section of the beam are denoted by

$$\boldsymbol{\beta} = \beta_x \mathbf{i} + \beta_y \mathbf{j} + \beta_z \mathbf{k},$$
$$\beta_x = \psi, \quad \beta_y = -w', \quad \beta_z = v'. \tag{9.2.12}$$

Usual boundary conditions at the ends of a beam may be the prescriptions of

$$u \text{ or } N, \quad \psi \text{ or } \mathcal{M}_x,$$
$$v \text{ or } Q_y, \quad \beta_z \text{ or } \mathcal{M}_z, \tag{9.2.13}$$
$$w \text{ or } Q_z, \quad \beta_y \text{ or } \mathcal{M}_y.$$

9.3 A Point Magnet in a Beam

Consider a point magnet at an interior point with $x = a$ in an elastic beam. The free body diagram of the magnet is shown in Fig. 9.3. We are interested in the motion of the magnet under an applied magnetic induction $\mathbf{B}$. The magnetic field produced by the magnet is neglected. At the two cross-sections of the beam to the left and right of the magnet, there are extensional and shear forces as well as bending and twisting moments. The magnetic force $\mathbf{f}^M$ and couple $\mathbf{c}^M$ per unit volume of the magnet by the external magnetic induction $\mathbf{B}$ act on the spin continuum of the magnet. We have

$$\mathbf{f}^M = \mathbf{M} \cdot (\mathbf{B}\nabla), \quad \mathbf{c}^M = \mathbf{M} \times \mathbf{B}. \tag{9.3.1}$$

The linear and angular momentum equations of the magnet including both the lattice and spin continua are

$$N(a^+) - N(a^-) + f_x^M V = m\ddot{u},$$
$$Q_y(a^+) - Q_y(a^-) + f_y^M V = m\ddot{v}, \tag{9.3.2}$$
$$Q_z(a^+) - Q_z(a^-) + f_z^M V = m\ddot{w},$$

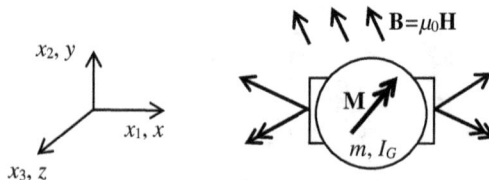

Fig. 9.3. Free body diagram of a magnet in a beam.

$$\mathcal{M}_x(a^+) - \mathcal{M}_x(a^-) + c_x^M V = \dot{L}_x,$$
$$\mathcal{M}_y(a^+) - \mathcal{M}_y(a^-) + c_y^M V = \dot{L}_y, \qquad (9.3.3)$$
$$\mathcal{M}_z(a^+) - \mathcal{M}_z(a^-) + c_z^M V = \dot{L}_z,$$

where V is the volume of the magnet. $\mathbf{L}$ is the total angular momentum of the magnet including contributions from both the lattice and the spin. For the beam equations in the previous section, Eqs. (9.3.2) and (9.3.3) serve as the jump conditions across a point magnet at $x = a$.

The small rotation and angular velocity of the lattice of the magnet are the same as those of the beam cross-section in Eq. (9.2.12) at $x = a$, i.e.,

$$\delta\boldsymbol{\theta}^l = \boldsymbol{\beta}|_{x=a}, \quad \boldsymbol{\omega}^l = \dot{\boldsymbol{\beta}}|_{x=a}. \qquad (9.3.4)$$

The total linear momentum of the magnet is from the lattice only. The total angular momentum of the magnet has contributions from both the lattice and the spin:

$$\mathbf{L} = I_G \boldsymbol{\omega}^l + \frac{1}{\gamma}\mathbf{M}V = I_G \dot{\boldsymbol{\beta}}|_{x=a} + \frac{1}{\gamma}\mathbf{M}V. \qquad (9.3.5)$$

The time rate of change of $\mathbf{L}$ is given by

$$\dot{\mathbf{L}} = I_G \dot{\boldsymbol{\omega}}^l + \frac{1}{\gamma}\dot{\mathbf{M}}V = I_G \ddot{\boldsymbol{\beta}}|_{x=a} + \frac{1}{\gamma}\dot{\mathbf{M}}V. \qquad (9.3.6)$$

$\dot{\mathbf{M}}$ is governed by the Landau–Lifshitz equation:

$$\frac{1}{\gamma}\dot{\mathbf{M}} = \mathbf{M} \times \mathbf{B}^{\mathrm{eff}} = \mathbf{M} \times (\mathbf{B} + \mathbf{B}^L), \qquad (9.3.7)$$

where $\mathbf{B} = \mathbf{B}^M$ and, from Eq. (9.1.11),

$$\mathbf{B}^L = -\lambda(\delta\mathbf{M})^{s/l}. \qquad (9.3.8)$$

Under the motion of the beam and the applied magnetic induction $\mathbf{B}$ which is assumed to be small and known, the magnetization $\mathbf{M}$ differs from its finite reference state $\mathbf{M}^0$ by a small amount of $\mathbf{m} = \delta\mathbf{M}$:

$$\mathbf{M} = \mathbf{M}^0 + \delta\mathbf{M} = \mathbf{M}^0 + \mathbf{m},$$
$$\mathbf{m} = \delta\mathbf{M} = \boldsymbol{\beta}|_{x=a} \times \mathbf{M}^0 + \delta\mathbf{M}^{s/l}. \qquad (9.3.9)$$

Since **m** and **B** are small, we have, approximately,

$$\mathbf{f}^M \cong \mathbf{M}^0 \cdot (\mathbf{B}\nabla), \quad \mathbf{c}^M \cong \mathbf{M}^0 \times \mathbf{B}, \tag{9.3.10}$$

which are known. The total angular momentum of the magnet may be written as

$$\mathbf{L} = I_G \dot{\boldsymbol{\beta}}|_{x=a} + \frac{1}{\gamma}(\mathbf{M}^0 + \mathbf{m})V. \tag{9.3.11}$$

The time rate of change of **L** is

$$\dot{\mathbf{L}} = I_G \ddot{\boldsymbol{\beta}}|_{x=a} + \frac{1}{\gamma}\dot{\mathbf{m}}V. \tag{9.3.12}$$

$\dot{\mathbf{m}}$ is governed by the Landau–Lifshitz equation in Eq. (9.3.7) which is approximated by

$$\frac{1}{\gamma}\dot{\mathbf{m}} \cong \mathbf{M}^0 \times (\mathbf{B} + \mathbf{B}^L), \tag{9.3.13}$$

where

$$\mathbf{B}^L = -\lambda(\delta\mathbf{M})^{s/l} = -\lambda(\mathbf{m} - \boldsymbol{\beta}|_{x=a} \times \mathbf{M}^0). \tag{9.3.14}$$

Substituting Eq. (9.3.14) into Eq. (9.3.13), we obtain

$$\frac{1}{\gamma}\dot{\mathbf{m}} \cong \mathbf{M}^0 \times [\mathbf{B} - \lambda(\mathbf{m} - \boldsymbol{\beta}|_{x=a} \times \mathbf{M}^0)]. \tag{9.3.15}$$

In summary, the $\dot{\mathbf{L}}$ in the jump condition in Eq. (9.3.3) is given by Eq. (9.3.12) where **m** is governed by Eq. (9.3.15).

In the special case when the spin is fixed to the lattice, $\lambda = \infty$, $\delta\mathbf{M}^{s/l} = 0$, **m** has the following expression:

$$\mathbf{m} = \boldsymbol{\beta}|_{x=a} \times \mathbf{M}^0, \tag{9.3.16}$$

and, from Eq. (9.3.12),

$$\dot{\mathbf{L}} = I_G \ddot{\boldsymbol{\beta}}|_{x=a} + \frac{1}{\gamma}\dot{\mathbf{m}}V = I_G \ddot{\boldsymbol{\beta}}|_{x=a} + \frac{1}{\gamma}\dot{\boldsymbol{\beta}}|_{x=a} \times \mathbf{M}^0 V. \tag{9.3.17}$$

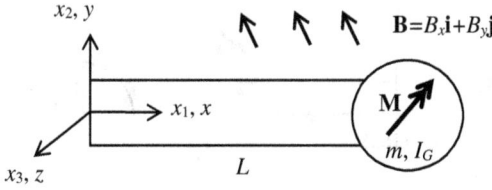

Fig. 9.4. An elastic cantilever with an end magnet in a magnetic field in the (x, y) plane.

9.4 Bending of an Elastic Beam with a Magnet

Consider the elastic cantilever in Fig. 9.4 with a fixed left end and a point magnet at its right end. The spin of the magnet is fixed to its lattice. The beam and the initial $\mathbf{M}^0$ define a plane which is chosen to be the (x, y) plane. The applied $\mathbf{B}$ is assumed to be in the (x, y) plane too. $\mathbf{B}$ is also assumed to be static and uniform. In this case $\mathbf{M}^0$ experiences a couple only, i.e.,

$$\mathbf{f}^M = \mathbf{M} \cdot (\mathbf{B}\nabla) = 0,$$

$$\mathbf{c}^M = \mathbf{M} \times \mathbf{B} \cong \mathbf{M}^0 \times \mathbf{B} = (M_x^0 \mathbf{i} + M_y^0 \mathbf{j}) \times (B_x \mathbf{i} + B_y \mathbf{j}) \qquad (9.4.1)$$

$$= (M_x^0 B_y - M_y^0 B_x)\mathbf{k} = c_z^M \mathbf{k}.$$

In this case, the end magnet is equivalent to an end bending moment. Hence, the beam is in bending in the (x, y) plane with a deflection curve described by $v(x)$. For static bending, the boundary-value problem is

$$-EIv'''' = 0, \quad 0 < x < L,$$

$$v(0) = 0,$$

$$\beta_z(0) = v'(0) = 0, \qquad (9.4.2)$$

$$Q_y(L) = -EIv'''(L) = 0,$$

$$-\mathcal{M}_z(L^-) + c_z^M V = -EIv''(L^-) + (M_x^0 B_y - M_y^0 B_x)V = 0.$$

The beam is in pure bending with a constant curvature. The solution for the deflection curve from Eq. (9.4.2) is

$$v(x) = \frac{x^2}{2EI}(M_x^0 B_y - M_y^0 B_x)V. \qquad (9.4.3)$$

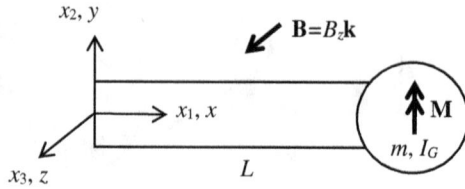

Fig. 9.5. An elastic cantilever with an end magnet in a magnetic field along the z-axis.

9.5 Torsion of an Elastic Beam with a Magnet

Consider the elastic cantilever with an end magnet in Fig. 9.5. The initial $\mathbf{M}^0 = M_y^0 \mathbf{j}$. The applied magnetic induction is $\mathbf{B} = B_z \mathbf{k}$ which is static and uniform. The magnetic loads on $\mathbf{M}^0$ are given by

$$\mathbf{f}^M = \mathbf{M} \cdot (\mathbf{B}\nabla) = 0,$$
$$\mathbf{c}^M = \mathbf{M} \times \mathbf{B} \cong \mathbf{M}^0 \times \mathbf{B} = M_y^0 \mathbf{j} \times B_z \mathbf{k} = M_y^0 B_z \mathbf{i} = c_x^M \mathbf{i}. \tag{9.5.1}$$

In this case, the end magnet is equivalent to an end twisting moment. The beam is in static torsion which is governed by

$$GI_p \psi'' = 0, \quad 0 < x < L,$$
$$\beta_x(0) = \psi(0) = 0, \tag{9.5.2}$$
$$-\mathcal{M}_x(L^-) + c_x^M V = -GI_p \psi'(L) + M_y^0 B_z V = 0.$$

The solution for the angle of twist is

$$\psi(x) = \frac{M_y^0 B_z V}{GI_p} x. \tag{9.5.3}$$

9.6 Elastically Connected Magnets

Elastically interacting magnets are common structures in devices [44, 45]. Figure 9.6 shows a simple example. The magnets are sufficiently far away from each other. Their magnetic interactions are neglected. The applied $\mathbf{B}$ is static and uniform. The basic approach of analyzing a beam with an interior concentrated load is to break the beam into two parts at $x = a$, solve the governing equations in each part, impose boundary conditions at both ends, and apply continuity/jump conditions at $x = a$ [48]. Since the problem is

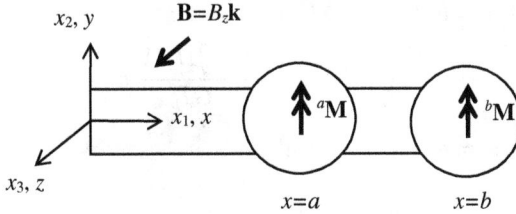

Fig. 9.6. A beam with two magnets in torsion.

linear, we use the method of superposition based on the solution of the uniform torsion problem in the previous section. The twisting moments at $x = a$ and b are

$$
\begin{aligned}
{}^a\mathbf{c}^M &\cong {}^aM_y^0\mathbf{j} \times B_z\mathbf{k} = {}^aM_y^0B_z\mathbf{i} = {}^ac_x^M\mathbf{i}, \\
{}^b\mathbf{c}^M &\cong {}^bM_y^0\mathbf{j} \times B_z\mathbf{k} = {}^bM_y^0B_z\mathbf{i} = {}^bc_x^M\mathbf{i}.
\end{aligned}
\tag{9.6.1}
$$

When the two magnets exist separately, we have

$$
\psi(x, {}^a\mathbf{M}) = \begin{cases} \dfrac{{}^aM_y^0B_z{}^aV}{GI_p}x, & 0 < x < a, \\[3mm] \dfrac{{}^aM_y^0B_z{}^aV}{GI_p}a, & a < x < b, \end{cases}
\tag{9.6.2}
$$

$$
\psi(x, {}^b\mathbf{M}) = \frac{{}^bM_y^0B_z{}^bV}{GI_p}x, \quad 0 < x < b.
\tag{9.6.3}
$$

When the two magnets are both present, we add Eqs. (9.6.2) and (9.6.3) to obtain

$$
\psi(x, {}^a\mathbf{M}, {}^b\mathbf{M}) = \psi(x, {}^a\mathbf{M}) + \psi(x, {}^b\mathbf{M})
$$

$$
= \begin{cases} \dfrac{{}^aM_y^0B_z{}^aV}{GI_p}x + \dfrac{{}^bM_y^0B_z{}^bV}{GI_p}x, & 0 < x < a, \\[3mm] \dfrac{{}^aM_y^0B_z{}^aV}{GI_p}a + \dfrac{{}^bM_y^0B_z{}^bV}{GI_p}x, & a < x < b. \end{cases}
\tag{9.6.4}
$$

The rotation at $x = b$ due to $^a\mathbf{M}$ alone is

$$\psi(b, {}^a\mathbf{M}, 0) = \frac{{}^aM_y^0 B_z\,{}^aV}{GI_p}a. \tag{9.6.5}$$

The rotation at $x = a$ due to $^b\mathbf{M}$ alone is

$$\psi(a, 0, {}^b\mathbf{M}) = \frac{{}^bM_y^0 B_z\,{}^bV}{GI_p}a. \tag{9.6.6}$$

We note that

$$\frac{\psi(b, {}^a\mathbf{M}, 0)}{{}^aM_y^0 B_z\,{}^aV} = \frac{a}{GI_p} = \frac{\psi(a, 0, {}^b\mathbf{M})}{{}^bM_y^0 B_z\,{}^bV}, \tag{9.6.7}$$

which is known as reciprocity.

9.7 Bending of a Piezoelectric Beam with a Magnet

Consider the piezoelectric beam in Fig. 9.7. It has two layers of polarized ceramics with opposite poling directions and is called a bimorph [38, 48]. When the poling direction is reversed, the elastic and dielectric constants remain the same but the piezoelectric constants change their signs. Under a voltage $\mathcal{V}(t)$ across the electrodes at the top and bottom of the bimorph, one piezoelectric layer elongates and the other contracts along x or vice versa. Hence, bending in the (x, z) plane is produced piezoelectrically. The beam has an end magnet whose $\mathbf{M}^0$ and the applied $\mathbf{B}$ are both in the (x, z) plane. We consider the case when $\mathcal{V}$ and $\mathbf{B}$ are static and $\mathbf{B}$ is uniform.

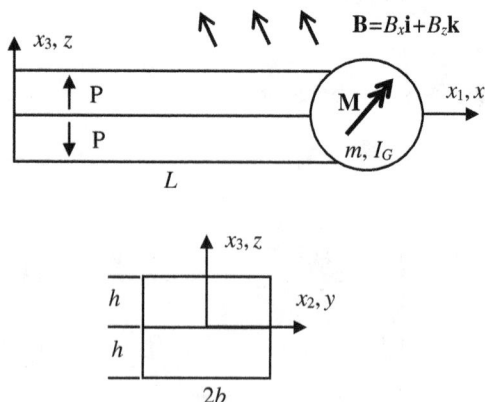

Fig. 9.7. A piezoelectric cantilever and its cross-section.

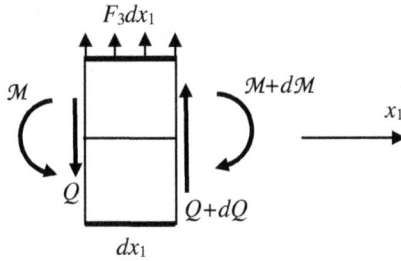

Fig. 9.8. A differential element of the beam and mechanical loads.

The deflection of the beam is described by

$$u_3 \cong u_3(x_1, t). \tag{9.7.1}$$

The bending moment $\mathcal{M} = \mathcal{M}_y$, shear force $\mathcal{Q} = \mathcal{Q}_z$ and the transvers load F_3 per unit length of the beam are shown in Fig. 9.8. The expressions of $\mathcal{M}$ and $\mathcal{Q}$ are [38, 48]

$$\mathcal{M} = \int_A T_{11} x_3 dA = -\bar{c}_{11} I u_{3,11} + \bar{e}_{31} bh \mathcal{V}, \tag{9.7.2}$$

$$\mathcal{Q} = \frac{\partial \mathcal{M}}{\partial x_1} = -\bar{c}_{11} I u_{3,111}, \tag{9.7.3}$$

where $\bar{c}_{11}$ and $\bar{e}_{31}$ are the effective one-dimensional elastic and piezoelectric constants:

$$\bar{c}_{11} = \frac{1}{s_{11}}, \quad \bar{e}_{31} = \frac{d_{31}}{s_{11}}, \quad A = 4bh, \quad I = \frac{4bh^3}{3}. \tag{9.7.4}$$

d_{31} is the relevant three-dimensional piezoelectric constant in the matrix notation [38, 48]. The equation for the flexural motion of the beam is obtained by applying Newton's second law to the differential element in Fig. 9.8 in the x_3 direction:

$$\frac{\partial \mathcal{Q}}{\partial x_1} + F_3 = \rho A \ddot{u}_3. \tag{9.7.5}$$

The substitution of Eq. (9.7.3) into Eq. (9.7.5) yields an equation for u_3:

$$-I\bar{c}_{11}u_{3,1111} + F_3 = \rho A\ddot{u}_3. \tag{9.7.6}$$

The magnetic loads on $\mathbf{M}^0$ are given by

$$\mathbf{f}^M = \mathbf{M} \cdot (\mathbf{B}\nabla) = 0,$$

$$\mathbf{c}^M = \mathbf{M} \times \mathbf{B} \cong \mathbf{M}^0 \times \mathbf{B} \tag{9.7.7}$$

$$= (M_x^0\mathbf{i} + M_z^0\mathbf{k}) \times (B_x\mathbf{i} + B_z\mathbf{k}) = (M_z^0 B_x - M_x^0 B_z)\mathbf{j} = c_y^M\mathbf{j}.$$

The boundary-value problem for the beam in static bending under $\mathcal{V}$ and $\mathbf{B}$ is

$$-\bar{c}_{11}Iu_{3,1111} = 0, \quad 0 < x < L, \tag{9.7.8}$$

$$u_3(0) = 0,$$

$$\beta_y(0) = -u_{3,1}(0) = 0, \tag{9.7.9}$$

$$\mathcal{Q}_z(L) = -\bar{c}_{11}Iu_{3,111}(L) = 0,$$

$$-\mathcal{M}_y(L^-) + c_y^M V = \bar{c}_{11}Iu_{3,11}(L^-) - \bar{e}_{31}bh\mathcal{V} \tag{9.7.10}$$

$$+ (M_z^0 B_x - M_x^0 B_z)V = 0,$$

where Eq. (9.3.3)$_2$ has been used. It can be seen that $\mathcal{V}$ and $\mathbf{B}$ produce an effective bending moment at the right end. The solution for the deflection curve is

$$u_3(x) = \frac{x^2}{2\bar{c}_{11}I}[\bar{e}_{31}bh\mathcal{V} - (M_z^0 B_x - M_x^0 B_z)V]. \tag{9.7.11}$$

9.8   Torsion of a Piezoelectric Beam with a Magnet

Consider the circular cylindrical beam of inner radius a and outer radius b shown in Fig. 9.9. The cross-section is denoted by A. The cylinder is made of polarized ceramics with circumferential poling along θ. We choose (r, θ, z) to correspond to indices $(2, 3, 1)$ so that the poling direction corresponds to 3. The lateral cylindrical surfaces of the beam are traction free and are unelectroded. The electric field in the free-space is neglected as an approximation. $\mathbf{B}$ is static and uniform.

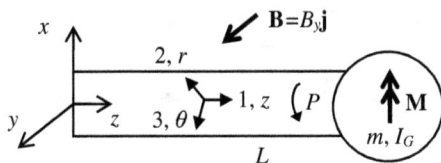

Fig. 9.9. A piezoelectric cantilever with an end magnet in a magnetic field along the y axis.

The magnetic loads on $\mathbf{M}^0$ are given by

$$\mathbf{f}^M = \mathbf{M} \cdot (\mathbf{B}\nabla) = 0,$$
$$\mathbf{c}^M = \mathbf{M} \times \mathbf{B} \cong \mathbf{M}^0 \times \mathbf{B} = M_x^0 \mathbf{i} \times B_y \mathbf{j} = M_x^0 B_y \mathbf{k} = c_z^M \mathbf{k}. \qquad (9.8.1)$$

The angle of twist and the electric potential are approximated by functions of the axial coordinate z and time only [38, 48]:

$$\psi = \psi(z,t), \quad \varphi = \varphi(z,t). \qquad (9.8.2)$$

The twisting moment $\mathcal{M} = \mathcal{M}_z$ and the axial electric displacement $\hat{D}_z$ over a cross-section are given by [38, 48]:

$$\mathcal{M} = \mathcal{M}_z = \int_A T_{z\theta} r \, dA = c_{55} I_p \psi_{,z} + e_{15} C \varphi_{,z}, \qquad (9.8.3)$$

$$\hat{D}_z = \int_A D_z \, dA = e_{15} C \psi_{,z} - \varepsilon_{11} A \varphi_{,z}, \qquad (9.8.4)$$

where c_{55} is the relevant shear elastic constant, e_{15} the relevant piezoelectric constant, ε_{11} the relevant dielectric constant, and

$$A = \pi(b^2 - a^2), \quad C = \frac{2\pi}{3}(b^3 - a^3), \quad I_p = \frac{\pi}{2}(b^4 - a^4). \qquad (9.8.5)$$

The equation for torsional motion is obtained by the moment equation of the differential element in Fig. 9.10:

$$\mathcal{M}_{,z} + m_z = \rho I_p \ddot{\psi}, \qquad (9.8.6)$$

where $m_z(z,t)$ is the distributed torsional load per unit length of the beam. Similarly, the one-dimensional charge equation of electrostatics can be obtained by applying electric loads to the differential element in Fig. 9.10 [38, 48]:

$$\hat{D}_{z,z} = 0. \qquad (9.8.7)$$

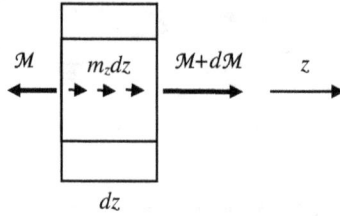

Fig. 9.10. A differential element of the beam under torsional loads.

Substituting Eqs. (9.8.3) and (9.8.4) into Eqs. (9.8.6) and (9.8.7), for a uniform beam, we arrive at two equations for ψ and φ as

$$c_{55}I_p\psi_{,zz} + e_{15}C\varphi_{,zz} + m_z = \rho I_p\ddot{\psi},$$
$$e_{15}C\psi_{,zz} - \varepsilon_{11}A\varphi_{,zz} = 0.$$

(9.8.8)

For the beam in static torsion under **B** and a constant voltage $\mathcal{V}$ across the two ends of the beam, the boundary-value problem is

$$c_{55}I_p\psi_{,zz} + e_{15}C\varphi_{,zz} = 0, \quad 0 < z < L,$$
$$e_{15}C\psi_{,zz} - \varepsilon_{11}A\varphi_{,zz} = 0, \quad 0 < z < L.$$

(9.8.9)

$$\varphi(0) = 0,$$
$$\beta_z(0) = \psi(0) = 0,$$

(9.8.10)

$$\varphi(L) = \mathcal{V},$$
$$-\mathcal{M}_z(L^-) + c_z^M V = -[c_{55}I_p\psi_{,z}(L) + e_{15}C\varphi_{,z}(L)] + M_x^0 B_y V = 0,$$

(9.8.11)

where the left end is fixed mechanically and grounded electrically, and Eq. (9.3.3)$_3$ has been used. It can be seen that $\mathcal{V}$ and **B** produce an effective twisting moment at the right end. The solution for the electric potential φ and the angle of twist ψ is

$$\varphi(x) = \frac{\mathcal{V}}{L}z,$$

(9.8.12)

$$\psi(x) = \frac{z}{c_{55}I_p}\left[-e_{15}C\frac{\mathcal{V}}{L} + M_x^0 B_y V\right].$$

Chapter 10

FERROMAGNETOELASTIC
CONDUCTORS

So far we have considered ferromagnetic insulators only. Electrical conduction in ferromagnets is a relevant topic. In this chapter, we develop a macroscopic theory of elastic and saturated ferromagnetic conductors using a four-continuum model [49]. The complete set of Maxwell's equations for electromagnetic fields or waves is used. The equations obtained in this chapter can describe interactions of elastic, electromagnetic and spin waves or phonon–photon–magnon couplings.

10.1   Four-Continuum Model

The basic behaviors of elastic ferromagnetic conductors can be modeled by the four interpenetrating and interacting continua in Figs. 10.1–10.3 and 10.5. The four continua are discussed in the following one at a time.

In the reference state without any deformation and fields at t_0, the reference position of a material point of the lattice continuum in Fig. 10.1 is $\mathbf{X}$. The reference mass density and charge density of the lattice continuum are $\rho^0(\mathbf{X})$ and $_0\sigma^l(\mathbf{X})$. At time t, the lattice continuum occupies a region v with a boundary surface s whose outward unit normal is $\mathbf{n}$. The current position, mass density and charge density of the material point associated with $\mathbf{X}$ are $\mathbf{y}$, ρ and σ^l, respectively. The motion of the lattice continuum is described by $\mathbf{y} = \mathbf{y}(\mathbf{X}, t)$. The velocity field of the lattice continuum is denoted by $\mathbf{v}(\mathbf{y}, t)$. We have the following equations from the conservation of mass and charge for the lattice continuum:

$$\frac{d\rho}{dt} + \rho v_{k,k} = 0, \qquad (10.1.1)$$

$$\frac{d\sigma^l}{dt} + \sigma^l v_{k,k} = 0. \qquad (10.1.2)$$

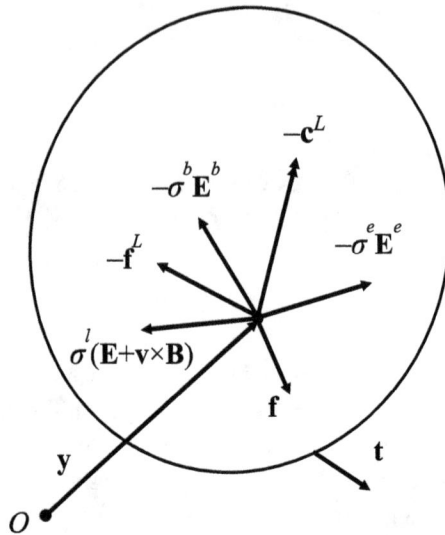

Fig. 10.1. Current configuration of the lattice continuum.

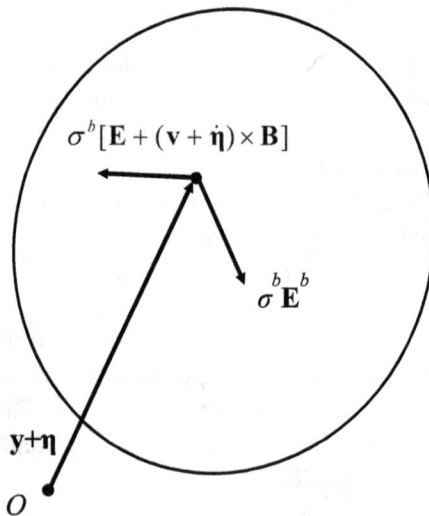

Fig. 10.2. Current configuration of the bound charge continuum.

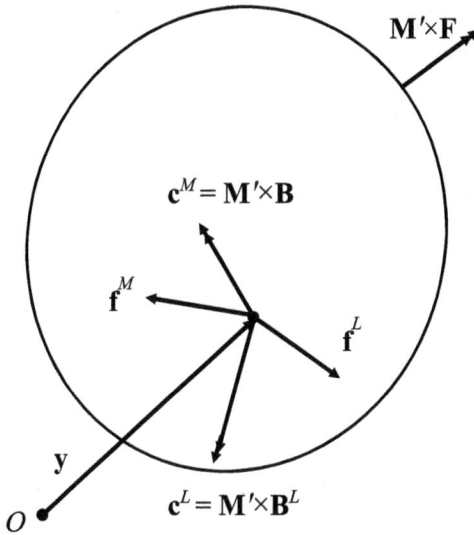

Fig. 10.3. Current configuration of the spin continuum.

The lattice continuum is under the usual mechanical surface traction $\mathbf{t}$ and mechanical body force $\mathbf{f}$. $\sigma^l(\mathbf{y})$ experiences a force under the Maxwellian electric field $\mathbf{E}$ and magnetic induction $\mathbf{B}$. The lattice continuum interacts with the bound charge continuum in Fig. 10.2 with an effective electric field $\mathbf{E}^b$. The lattice continuum interacts with the spin continuum in Fig. 10.3 with a local force $\mathbf{f}^L$ and a local couple $\mathbf{c}^L$ produced by an effective local magnetic induction $\mathbf{B}^L$. The lattice continuum also interacts with the free charge fluid in Fig. 10.5 with an effective electric field $\mathbf{E}^e$.

The bound charge continuum [7] in Fig. 10.2 is massless and has a negative charge density $_0\sigma^b(\mathbf{X})$ at the reference state. The residual charge density $\sigma^r(\mathbf{X})$ at the reference state is defined by

$$_0\sigma^l(\mathbf{X}) + {_0\sigma^b}(\mathbf{X}) = \sigma^r(\mathbf{X}). \tag{10.1.3}$$

The bound charge continuum can displace a little from the lattice continuum by an infinitesimal displacement field $\boldsymbol{\eta}(\mathbf{y}, t)$. $\boldsymbol{\eta}$ is assumed to preserve the volume of the bound charge continuum, i.e.,

$$\eta_{k,k} = \frac{\partial \eta_k}{\partial y_k} = \nabla \cdot \boldsymbol{\eta} = 0. \tag{10.1.4}$$

In the current state, $\sigma^b(\mathbf{y} + \boldsymbol{\eta})$ experiences a force under the Maxwellian electric field $\mathbf{E}$ and magnetic induction $\mathbf{B}$. We also have

$$\sigma^l(\mathbf{y}, t) + \sigma^b(\mathbf{y} + \boldsymbol{\eta}, t) = \sigma^r(\mathbf{y}, t). \tag{10.1.5}$$

The bound charge and the residual charge are conserved, i.e.,

$$\frac{d\sigma^b}{dt} + \sigma^b(v_k + \dot{\eta}_k)_{,k} = \frac{d\sigma^b}{dt} + \sigma^b v_{k,k} = 0, \tag{10.1.6}$$

$$\frac{d\sigma^r}{dt} + \sigma^r v_{k,k} = 0. \tag{10.1.7}$$

From Eqs. (10.1.1), (10.1.2) and (10.1.6), we obtain

$$\frac{\dot{\sigma}^l}{\sigma^l} = \frac{\dot{\sigma}^b}{\sigma^b} = \frac{\dot{\rho}}{\rho} = -v_{k,k}. \tag{10.1.8}$$

In terms of $\boldsymbol{\eta}$, the polarization per unit volume is defined by [7]

$$\mathbf{P} = \sigma^l(\mathbf{y})(-\boldsymbol{\eta}) = \sigma^b(\mathbf{y} + \boldsymbol{\eta})\boldsymbol{\eta} \cong \sigma^b(\mathbf{y})\boldsymbol{\eta}. \tag{10.1.9}$$

For later use, we introduce the polarization per unit mass by

$$\boldsymbol{\pi} = \frac{\mathbf{P}}{\rho}. \tag{10.1.10}$$

We have

$$\sigma^b \dot{\boldsymbol{\eta}} = \frac{d}{dt}(\sigma^b \boldsymbol{\eta}) - \dot{\sigma}^b \boldsymbol{\eta} = \dot{\mathbf{P}} - \frac{\mathbf{P}}{\sigma^b}\dot{\sigma}^b$$

$$= \dot{\mathbf{P}} - \mathbf{P}\frac{\dot{\rho}}{\rho} = \rho\dot{\boldsymbol{\pi}} + \dot{\rho}\boldsymbol{\pi} - \dot{\rho}\boldsymbol{\pi} = \rho\dot{\boldsymbol{\pi}}. \tag{10.1.11}$$

The spin continuum [16] in Fig. 10.3 is massless. It cannot displace with respect to the lattice continuum. The spin continuum carries distributed circulating currents which produce a magnetic moment field $\mathbf{M}'(\mathbf{y}, t)$ per unit volume in the co-moving or instantaneous rest frame of the material point of the lattice at $\mathbf{y}$ [15]. $\mathbf{M}'$ can rotate with respect to the lattice. It experiences a force $\mathbf{f}^M$ and a couple $\mathbf{c}^M$ under the Maxwellian magnetic induction $\mathbf{B}$. We have

$$\mathbf{f}^M = \mathbf{M}' \cdot (\mathbf{B}\nabla), \quad \mathbf{c}^M = \mathbf{M}' \times \mathbf{B}, \quad \mathbf{c}^L = \mathbf{M}' \times \mathbf{B}^L. \tag{10.1.12}$$

The spin continuum also experiences a distributed couple $\mathbf{M}' \times \mathbf{F}$ per unit area on its boundary surface due to an effective exchange field $\mathbf{F}$ [15]. Since

$\mathbf{F}$ and $\mathbf{B}^L$ act on $\mathbf{M}'$ through cross products, the components of $\mathbf{F}$ and $\mathbf{B}^L$ along $\mathbf{M}'$ have no contributions. Hence, it can be assumed that [15]

$$\mathbf{F} \cdot \mathbf{M}' = 0, \quad \mathbf{B}^L \cdot \mathbf{M}' = 0. \tag{10.1.13}$$

For later use, we introduce the magnetization per unit mass in the co-moving reference frame by

$$\boldsymbol{\mu}' = \frac{\mathbf{M}'}{\rho}. \tag{10.1.14}$$

The saturation condition is

$$\boldsymbol{\mu}' \cdot \boldsymbol{\mu}' = \mu_s'^2, \tag{10.1.15}$$

where μ_s' is a constant (saturation magnetization) and s is not a tensor index. Mathematically, Eq. (10.1.15) is a constraint on $\boldsymbol{\mu}'$. With differentiations with respect to t and/or $\mathbf{X}$, Eq. (10.1.15) implies that

$$\mu_k' \frac{d\mu_k'}{dt} = 0, \quad \mu_k'\mu_{k,L}' = 0, \quad \mu_{k,L}' \frac{d\mu_k'}{dt} + \mu_k' \frac{d\mu_{k,L}'}{dt} = 0. \tag{10.1.16}$$

Although the magnitude of $\boldsymbol{\mu}'$ cannot change because of the saturation condition, $\boldsymbol{\mu}'$ can still change its direction as described by the angular displacement $\delta\theta = |\delta\boldsymbol{\mu}'|/|\boldsymbol{\mu}'|$ in Fig. 10.4. We introduce an angular displacement vector $\boldsymbol{\delta\theta}$ by

$$\boldsymbol{\delta\theta} = \delta\theta \frac{\boldsymbol{\mu}'}{|\boldsymbol{\mu}'|} \times \frac{\delta\boldsymbol{\mu}'}{|\delta\boldsymbol{\mu}'|}$$

$$= \frac{|\delta\boldsymbol{\mu}'|}{|\boldsymbol{\mu}'|} \frac{\boldsymbol{\mu}'}{|\boldsymbol{\mu}'|} \times \frac{\delta\boldsymbol{\mu}'}{|\delta\boldsymbol{\mu}'|} = \frac{1}{\mu_s'^2} \boldsymbol{\mu}' \times (\delta\boldsymbol{\mu}'). \tag{10.1.17}$$

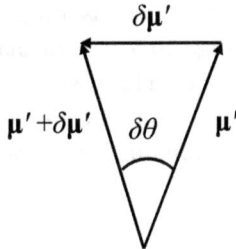

Fig. 10.4. Change of direction of $\boldsymbol{\mu}'$.

Then

$$\delta\boldsymbol{\theta} \times \boldsymbol{\mu}' = \frac{1}{\mu_s'^2}(\boldsymbol{\mu}' \times \delta\boldsymbol{\mu}') \times \boldsymbol{\mu}'$$

$$= \frac{1}{\mu_s'^2}[(\boldsymbol{\mu}' \cdot \boldsymbol{\mu}')\delta\boldsymbol{\mu}' - (\boldsymbol{\mu}' \cdot \delta\boldsymbol{\mu}')\boldsymbol{\mu}'] = \delta\boldsymbol{\mu}', \qquad (10.1.18)$$

where the following vector identity has been used [14]:

$$\mathbf{A} \times (\mathbf{B} \times \mathbf{C}) = (\mathbf{A} \cdot \mathbf{C})\mathbf{B} - (\mathbf{A} \cdot \mathbf{B})\mathbf{C}. \qquad (10.1.19)$$

We also introduce an angular velocity vector for a saturated $\boldsymbol{\mu}'$ through

$$\boldsymbol{\omega} = \lim_{\delta t \to 0} \frac{\delta\boldsymbol{\theta}}{\delta t} = \frac{1}{\mu_s'^2}\boldsymbol{\mu}' \times \frac{d\boldsymbol{\mu}'}{dt}. \qquad (10.1.20)$$

The power of a magnetic couple $\boldsymbol{\Gamma} = \rho\boldsymbol{\mu}' \times \mathbf{B}$ on a saturated magnetization is given by

$$\boldsymbol{\Gamma} \cdot \boldsymbol{\omega} = (\rho\boldsymbol{\mu}' \times \mathbf{B}) \cdot \left(\frac{1}{\mu_s'^2}\boldsymbol{\mu}' \times \frac{d\boldsymbol{\mu}'}{dt}\right)$$

$$= \frac{\rho}{\mu_s'^2}\left[(\boldsymbol{\mu}' \cdot \boldsymbol{\mu}')\left(\mathbf{B} \cdot \frac{d\boldsymbol{\mu}'}{dt}\right) - \left(\boldsymbol{\mu}' \cdot \frac{d\boldsymbol{\mu}'}{dt}\right)(\mathbf{B} \cdot \boldsymbol{\mu}')\right] = \rho\mathbf{B} \cdot \frac{d\boldsymbol{\mu}'}{dt},$$

$$(10.1.21)$$

where we have used the following vector identity [14]:

$$(\mathbf{u} \times \mathbf{v}) \cdot (\mathbf{w} \times \mathbf{t}) = (\mathbf{u} \cdot \mathbf{w})(\mathbf{v} \cdot \mathbf{t}) - (\mathbf{u} \cdot \mathbf{t})(\mathbf{v} \cdot \mathbf{w}). \qquad (10.1.22)$$

Following [15], we write the angular momentum of $\mathbf{M}'$ as

$$\mathbf{L}' = \frac{1}{\gamma}\mathbf{M}', \qquad (10.1.23)$$

where γ is the gyromagnetic ratio which is a negative number.

The free charge fluid in Fig. 10.5 is also massless. Its pressure field and internal energy are neglected. The free charge fluid is driven by the Maxwellian electric field $\mathbf{E}$ and magnetic induction $\mathbf{B}$, and experiences resistance from the lattice continuum. Its charge density is $\sigma^e(\mathbf{y}, t)$ and its velocity field is denoted by $\mathbf{v}^e(\mathbf{y}, t)$. The continuity equation for the free charge is

$$\frac{\partial\sigma^e}{\partial t} + (\sigma^e v_k^e)_{,k} = 0. \qquad (10.1.24)$$

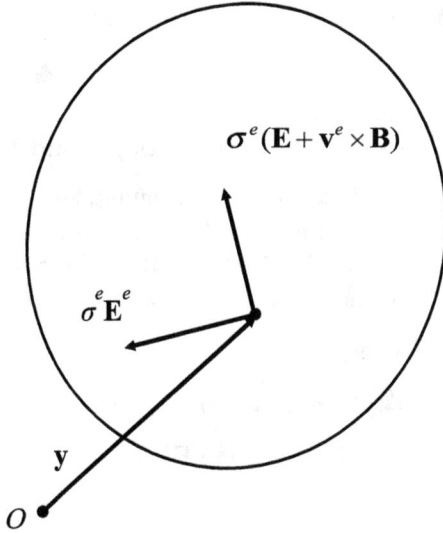

Fig. 10.5. Current configuration of the free charge fluid.

The charge and current densities are denoted by

$$\sigma = \sigma^r + \sigma^e, \quad \mathbf{J} = \sigma^r \mathbf{v} + \sigma^e \mathbf{v}^e, \tag{10.1.25}$$

which satisfy

$$\frac{\partial \sigma}{\partial t} + J_{k,k} = 0. \tag{10.1.26}$$

Gauss' law for the electric field can be written as

$$\oint_s \varepsilon_0 \mathbf{n} \cdot \mathbf{E} ds = \int_v [\sigma^l(\mathbf{y}) + \sigma^b(\mathbf{y}) + \sigma^e(\mathbf{y})] dv. \tag{10.1.27}$$

The differential form of (10.1.27) is

$$\begin{aligned}
\varepsilon_0 E_{k,k} &= \sigma^l(\mathbf{y}) + \sigma^b(\mathbf{y}) + \sigma^e(\mathbf{y}) \\
&= -\sigma^b(\mathbf{y}+\boldsymbol{\eta}) + \sigma^r(\mathbf{y}) + \sigma^b(\mathbf{y}) + \sigma^e(\mathbf{y}) \\
&= -\sigma^b(\mathbf{y}) - \sigma^b_{,i}(\mathbf{y})\eta_i + \sigma^r(\mathbf{y}) + \sigma^b(\mathbf{y}) + \sigma^e(\mathbf{y}) \\
&= -[\sigma^b(\mathbf{y})\eta_i]_{,i} + \sigma^b(\mathbf{y})\eta_{i,i} + \sigma^r(\mathbf{y}) + \sigma^e(\mathbf{y}) \\
&= -P_{i,i} + \sigma^r(\mathbf{y}) + \sigma^e(\mathbf{y}) = -P_{i,i} + \sigma.
\end{aligned} \tag{10.1.28}$$

Equation (10.1.28) can be written as

$$(\varepsilon_0 E_i + P_i)_{,i} = D_{i,i} = \sigma, \tag{10.1.29}$$

where we have introduced the electric displacement vector $\mathbf{D}$ through

$$D_i = \varepsilon_0 E_i + P_i. \tag{10.1.30}$$

10.2   Electromagnetic Body Force, Couple and Power

Consider a unit volume of the lattice continuum, spin continuum and free charge fluid at $\mathbf{y}$. The associated bound charge continuum at $\mathbf{y}+\boldsymbol{\eta}$ also has a unit volume. From Figs. 10.1–10.3 and 10.5, the electromagnetic body force on the above combined continuum of a unit volume is

$$
\begin{aligned}
\mathbf{F}^{\mathrm{EM}} = {} & \sigma^l(\mathbf{y})[\mathbf{E}(\mathbf{y}) + \mathbf{v}(\mathbf{y}) \times \mathbf{B}(\mathbf{y})] \\
& + \sigma^b(\mathbf{y}+\boldsymbol{\eta})[\mathbf{E}(\mathbf{y}+\boldsymbol{\eta}) + (\mathbf{v}+\dot{\boldsymbol{\eta}}) \times \mathbf{B}(\mathbf{y}+\boldsymbol{\eta})] \\
& + \mathbf{M}' \cdot (\mathbf{B}\nabla) + \sigma^e(\mathbf{y})[\mathbf{E}(\mathbf{y}) + \mathbf{v}^e(\mathbf{y}) \times \mathbf{B}(\mathbf{y})].
\end{aligned}
\tag{10.2.1}
$$

Using

$$
\begin{aligned}
E_j(\mathbf{y}+\boldsymbol{\eta}) &\cong E_j(\mathbf{y}) + \eta_i E_{j,i}(\mathbf{y}), \\
B_j(\mathbf{y}+\boldsymbol{\eta}) &\cong B_j(\mathbf{y}) + \eta_i B_{j,i}(\mathbf{y}),
\end{aligned}
\tag{10.2.2}
$$

we can write $\mathbf{F}^{\mathrm{EM}}$ as

$$
\mathbf{F}^{\mathrm{EM}} = \mathbf{P} \cdot \nabla \mathbf{E} + \mathbf{M}' \cdot (\mathbf{B}\nabla) + \mathbf{v} \times (\mathbf{P} \cdot \nabla \mathbf{B}) + \rho\dot{\boldsymbol{\pi}} \times \mathbf{B} + \sigma\mathbf{E} + \mathbf{J} \times \mathbf{B},
\tag{10.2.3}
$$

where a term of the product of the small $\boldsymbol{\eta}$, its material time derivative and the gradient of $\mathbf{B}$ has been omitted as an approximation [15]. It can be shown that [15]

$$
F_j^{\mathrm{EM}} = T_{ij,i}^{\mathrm{EM}} - \frac{\partial G_j}{\partial t},
\tag{10.2.4}
$$

where

$$
\mathbf{G} = \varepsilon_0 \mathbf{E} \times \mathbf{B}, \quad G_j = \varepsilon_0 \varepsilon_{jkl} E_k B_l
\tag{10.2.5}
$$

is the electromagnetic momentum density and

$$
\begin{aligned}
T_{ij}^{\mathrm{EM}} = {} & P_i E_j' - B_i M_j' + \varepsilon_0 E_i E_j + \frac{1}{\mu_0} B_i B_j \\
& - \frac{1}{2}\left(\varepsilon_0 E_k E_k + \frac{1}{\mu_0} B_k B_k - 2M_k' B_k \right) \delta_{ij}
\end{aligned}
\tag{10.2.6}
$$

is the electromagnetic stress tensor.

The electromagnetic couple on a unit volume of the combined continuum about $\mathbf{y} = 0$ is given by

$$
\begin{aligned}
\mathbf{y} \times [\sigma^l(\mathbf{y})\mathbf{E}(\mathbf{y}) &+ \sigma^l(\mathbf{y})\mathbf{v}(\mathbf{y}) \times \mathbf{B}(\mathbf{y}) + \mathbf{M}' \cdot (\mathbf{B}\nabla) \\
&+ \sigma^e(\mathbf{y})\mathbf{E}(\mathbf{y}) + \sigma^e(\mathbf{y})\mathbf{v}^e(\mathbf{y}) \times \mathbf{B}(\mathbf{y})] \\
&+ (\mathbf{y} + \boldsymbol{\eta}) \times [\sigma^b(\mathbf{y} + \boldsymbol{\eta})\mathbf{E}(\mathbf{y} + \boldsymbol{\eta}) \\
&+ \sigma^b(\mathbf{y} + \boldsymbol{\eta})(\mathbf{v} + \dot{\boldsymbol{\eta}}) \times \mathbf{B}(\mathbf{y} + \boldsymbol{\eta})] + \mathbf{M}' \times \mathbf{B} \\
&= \mathbf{y} \times \mathbf{F}^{\mathrm{EM}} + \mathbf{C}^{\mathrm{EM}},
\end{aligned}
\tag{10.2.7}
$$

where we have introduced the electromagnetic body couple $\mathbf{C}^{\mathrm{EM}}$ per unit volume by

$$
\mathbf{C}^{\mathrm{EM}} = \mathbf{P} \times \mathbf{E}' + \mathbf{M}' \times \mathbf{B},
\tag{10.2.8}
$$

$$
\mathbf{E}' = \mathbf{E} + \mathbf{v} \times \mathbf{B}, \quad \mathbf{M}' \cong \mathbf{M} + \mathbf{v} \times \mathbf{P}.
\tag{10.2.9}
$$

The electromagnetic power per unit volume of the combined continuum is

$$
\begin{aligned}
W^{\mathrm{EM}} &= \sigma^l(\mathbf{y})[\mathbf{E}(\mathbf{y}) + \mathbf{v}(\mathbf{y}) \times \mathbf{B}(\mathbf{y})] \cdot \mathbf{v} \\
&+ \sigma^e(\mathbf{y})[\mathbf{E}(\mathbf{y}) + \mathbf{v}^e(\mathbf{y}) \times \mathbf{B}(\mathbf{y})] \cdot \mathbf{v}^e \\
&+ \sigma^b(\mathbf{y} + \boldsymbol{\eta})[\mathbf{E}(\mathbf{y} + \boldsymbol{\eta}) + (\mathbf{v} + \dot{\boldsymbol{\eta}}) \times \mathbf{B}(\mathbf{y} + \boldsymbol{\eta})] \cdot (\mathbf{v} + \dot{\boldsymbol{\eta}}) - \mathbf{M}' \cdot \frac{\partial \mathbf{B}}{\partial t} \\
&= (\mathbf{P} \cdot \nabla \mathbf{E}) \cdot \mathbf{v} + \rho \mathbf{E} \cdot \dot{\boldsymbol{\pi}} - \mathbf{M}' \cdot \frac{\partial \mathbf{B}}{\partial t} + \mathbf{E} \cdot \mathbf{J} \\
&= \mathbf{F}^{\mathrm{EM}} \cdot \mathbf{v} + \rho \mathbf{E}' \cdot \dot{\boldsymbol{\pi}} - \mathbf{M}' \cdot \dot{\mathbf{B}} + \mathbf{J}' \cdot \mathbf{E}',
\end{aligned}
\tag{10.2.10}
$$

where

$$
\mathbf{J}' = \mathbf{J} - \sigma\mathbf{v} = \sigma^e(\mathbf{v}^e - \mathbf{v}).
\tag{10.2.11}
$$

10.3 Integral Balance Laws

The relevant balance laws from physics are as follows. For the electromagnetic fields, we have Maxwell's equations

$$
\oint_c \mathbf{E} \cdot d\mathbf{y} = -\frac{\partial}{\partial t} \int_s \mathbf{n} \cdot \mathbf{B} ds,
\tag{10.3.1}
$$

$$
\oint_c \mathbf{H} \cdot d\mathbf{y} = \frac{\partial}{\partial t} \int_s \mathbf{n} \cdot \mathbf{D} ds + \int_s \mathbf{n} \cdot \mathbf{J} ds,
\tag{10.3.2}
$$

$$\int_s \mathbf{n} \cdot \mathbf{D} ds = \int_v \sigma dv, \qquad (10.3.3)$$

$$\int_s \mathbf{n} \cdot \mathbf{B} ds = 0. \qquad (10.3.4)$$

The linear momentum equation and the angular momentum equation for the spin continuum are

$$\int_v (\mathbf{f}^M + \mathbf{f}^L) dv = 0, \qquad (10.3.5)$$

$$\frac{d}{dt} \int_v \rho \frac{\boldsymbol{\mu}'}{\gamma} dv = \int_v \mathbf{y} \times (\mathbf{f}^M + \mathbf{f}^L) dv$$

$$+ \int_s \rho \boldsymbol{\mu}' \times \mathbf{F} ds + \int_v \rho \boldsymbol{\mu}' \times (\mathbf{B} + \mathbf{B}^L) dv. \qquad (10.3.6)$$

The conservation of mass, the linear and angular momentum equations for the combined continuum are

$$\frac{d}{dt} \int_v \rho dv = 0, \qquad (10.3.7)$$

$$\frac{d}{dt} \int_v \rho \mathbf{v} dv = \int_s \mathbf{t} ds + \int_v (\rho \mathbf{f} + \mathbf{F}^{\mathrm{EM}}) dv, \qquad (10.3.8)$$

$$\frac{d}{dt} \int_v \left(\mathbf{y} \times \rho \mathbf{v} + \rho \frac{\boldsymbol{\mu}'}{\gamma} \right) dv$$

$$= \int_s \mathbf{y} \times \mathbf{t} ds + \int_s \rho \boldsymbol{\mu}' \times \mathbf{F} ds + \int_v [\mathbf{y} \times (\rho \mathbf{f} + \mathbf{F}^{\mathrm{EM}}) + \mathbf{C}^{\mathrm{EM}}] dv.$$

$$(10.3.9)$$

The energy equation and the second law of thermodynamics for the combined continuum are

$$\frac{d}{dt} \int_v \rho \left(\frac{1}{2} \mathbf{v} \cdot \mathbf{v} + \varepsilon \right) dv = \int_s \left(\mathbf{t} \cdot \mathbf{v} - \mathbf{n} \cdot \mathbf{q} + \mathbf{F} \cdot \rho \frac{d\boldsymbol{\mu}'}{dt} \right) ds$$

$$+ \int_v (\rho \mathbf{f} \cdot \mathbf{v} + \rho r) dv + \int_v W^{\mathrm{EM}} dv,$$

$$(10.3.10)$$

$$\frac{d}{dt} \int_v \rho \eta dv \geq \int_v \frac{\rho r}{\theta} dv - \int_s \frac{\mathbf{q} \cdot \mathbf{n}}{\theta} ds, \qquad (10.3.11)$$

where ε is the internal energy density per unit mass, r the body heat source per unit mass, $\mathbf{q}$ the heat flux vector, θ the absolute temperature and η the entropy density per unit mass. The kinetic energy associated with the spin angular momentum has been omitted since its material time derivative vanishes on account of Eqs. (10.3.7) and (10.1.15).

10.4 Differential Balance Laws

The differential forms of Eqs. (10.3.1)–(10.3.4) are

$$\nabla \cdot \mathbf{D} = \sigma, \tag{10.4.1}$$

$$\nabla \cdot \mathbf{B} = 0, \tag{10.4.2}$$

$$\nabla \times \mathbf{E} + \frac{\partial \mathbf{B}}{\partial t} = 0, \tag{10.4.3}$$

$$\nabla \times \mathbf{H} = \frac{\partial \mathbf{D}}{\partial t} + \mathbf{J}. \tag{10.4.4}$$

Since the spin continuum is massless, the linear momentum equation in Eq. (10.3.5) simply leads to $\mathbf{f}^M + \mathbf{f}^L = 0$. For the angular momentum equation in Eq. (10.3.6), we introduce an exchange tensor $\mathbf{A}$ by [15]

$$\mathbf{F} = -\mathbf{n} \cdot \mathbf{A}, \tag{10.4.5}$$

which is restricted by Eq. $(10.1.13)_1$ through

$$\mathbf{A} \cdot \boldsymbol{\mu}' = 0. \tag{10.4.6}$$

Then the angular momentum equation in Eq. (10.3.6) can be brought into the following differential form using the divergence theorem:

$$\frac{1}{\gamma} \rho \frac{d\mu_i'}{dt} = \varepsilon_{ijk} \rho \mu_j' \left(-A_{lk,l} - \frac{A_{lk}}{\rho} \rho_{,l} + B_k + B_k^L \right) - \varepsilon_{ijk} \rho A_{lk} \mu_{j,l}'. \tag{10.4.7}$$

As a cross product, the first term on the right-hand side of Eq. (10.4.7) is perpendicular to $\boldsymbol{\mu}'$. Dotting both sides of Eq. (10.4.7) by $\boldsymbol{\mu}'$, we have

$$\frac{1}{\gamma} \mu_i' \frac{d\mu_i'}{dt} = -\mu_i' \varepsilon_{ijk} A_{lk} \mu_{j,l}'. \tag{10.4.8}$$

The left-hand side of Eq. (10.4.8) vanishes because of the constraint due to saturation in Eq. $(10.1.16)_1$. Then Eq. (10.4.8) reduces to

$$-\mu_i' \varepsilon_{ijk} A_{lk} \mu_{j,l}' = 0. \tag{10.4.9}$$

To satisfy Eq. (10.4.9), we impose the following restriction on $\mathbf{A}$ [15]:

$$A_{lk} \mu_{j,l}' = A_{lj} \mu_{k,l}'. \tag{10.4.10}$$

With the use of Eq. (10.4.10), the angular momentum equation in Eq. (10.4.7) becomes

$$\frac{1}{\gamma}\frac{d\mu_i'}{dt} = \varepsilon_{ijk}\mu_j'\left(-A_{lk,l} - \frac{A_{lk}}{\rho}\rho_{,l} + B_k + B_k^L\right). \tag{10.4.11}$$

If we take a dot product of both sides of Eq. (10.4.11) with the following vector which is along the $\boldsymbol{\omega}$ in Eq. (10.1.20):

$$\varepsilon_{imn}\mu_m'\frac{d\mu_n'}{dt}, \tag{10.4.12}$$

we obtain

$$\frac{d\mu_k'}{dt}\left(-A_{lk,l} - \frac{A_{lk}}{\rho}\rho_{,l} + B_k + B_k^L\right) = 0, \tag{10.4.13}$$

which will be useful later. The conservation of mass in Eq. (10.3.7) leads to

$$\frac{d\rho}{dt} + \rho\nabla\cdot\mathbf{v} = 0. \tag{10.4.14}$$

The differential form of the linear momentum equation in Eq. (10.3.8) is

$$\rho\frac{d\mathbf{v}}{dt} = \nabla\cdot\boldsymbol{\tau} + \rho\mathbf{f} + \mathbf{F}^{\mathrm{EM}}. \tag{10.4.15}$$

From the angular momentum equation in Eq. (10.3.9), we obtain

$$\frac{d}{dt}\int_v \rho\frac{\boldsymbol{\mu}'}{\gamma}dv = \int_s \rho\boldsymbol{\mu}' \times \mathbf{F}ds + \int_v (\mathbf{e}_i\varepsilon_{ijk}\tau_{jk} + \mathbf{C}^{\mathrm{EM}})dv$$

$$+ \int_v \mathbf{y} \times (\nabla\cdot\boldsymbol{\tau} + \rho\mathbf{f} + \mathbf{F}^{\mathrm{EM}} - \rho\dot{\mathbf{v}})dv, \tag{10.4.16}$$

which results in

$$\frac{1}{\gamma}\rho\frac{d\mu_i'}{dt} = \varepsilon_{ijk}\tau_{jk} + C_i^{\mathrm{EM}} + \varepsilon_{ijk}\rho\mu_j'\left(-A_{lk,l} - \frac{A_{lk}}{\rho}\rho_{,l}\right). \tag{10.4.17}$$

From Eqs. (10.4.11) and (10.4.17), we obtain

$$\varepsilon_{ijk}\tau_{jk} + C_i^{\mathrm{EM}} - \varepsilon_{ijk}\rho\mu_j'(B_k + B_k^L) = 0, \tag{10.4.18}$$

or, after the use of Eq. (10.2.8),

$$\varepsilon_{ijk}\tau_{jk} + \varepsilon_{ijk}P_jE_k' - \varepsilon_{ijk}\rho\mu_j'B_k^L = 0. \tag{10.4.19}$$

The differential forms of Eqs. (10.3.10) and (10.3.11) can be obtained using the divergence theorem as

$$\rho\frac{d\varepsilon}{dt} = \tau_{ij}v_{j,i} - A_{ij,i}\rho\left(\frac{d\mu'_j}{dt}\right) - A_{ij}\rho_{,i}\left(\frac{d\mu'_j}{dt}\right) - A_{ij}\rho\left(\frac{d\mu'_j}{dt}\right)_{,i}$$

(10.4.20)

$$-\rho\mu'_i\frac{dB_i}{dt} + \rho E'_i\frac{d\pi_i}{dt} + J'_iE'_i + \rho r - q_{i,i},$$

(10.4.21)

$$\rho\frac{d\eta}{dt} \geq \frac{\rho r}{\theta} - \left(\frac{q_i}{\theta}\right)_{,i}.$$

10.5 Constitutive Relations

With the use of Eq. (10.4.13), the energy equation in Eq. (10.4.20) becomes

$$\rho\frac{d\varepsilon}{dt} = \tau_{ij}v_{j,i} - A_{ij}\rho\left(\frac{d\mu'_j}{dt}\right)_{,i} - (B_k + B_k^L)\rho\frac{d\mu'_k}{dt}$$

(10.5.1)

$$-\rho\mu'_i\frac{dB_i}{dt} + \rho E'_i\frac{d\pi_i}{dt} + J'_iE'_i + \rho r - q_{i,i}.$$

Under the following Legendre transform:

$$F = \varepsilon - E'_i\pi_i + B_i\mu'_i - \theta\eta,$$

(10.5.2)

Eq. (10.5.1) becomes

$$\rho\left(\frac{dF}{dt} + \frac{d\theta}{dt}\eta + \theta\frac{d\eta}{dt}\right) = \tau_{ij}v_{j,i} - A_{ij}\rho\left(\frac{d\mu'_j}{dt}\right)_{,i}$$

(10.5.3)

$$-B_k^L\rho\frac{d\mu'_k}{dt} - \rho\pi_i\frac{dE'_i}{dt} + J'_iE'_i + \rho r - q_{i,i}.$$

Eliminating r from Eqs. (10.4.21) and (10.5.3), we obtain the Clausius–Duhem inequality as

$$-\rho\left(\frac{dF}{dt} + \frac{d\theta}{dt}\eta\right) + \tau_{ij}v_{j,i} - A_{ij}\rho\left(\frac{d\mu'_j}{dt}\right)_{,i}$$

(10.5.4)

$$-B_k^L\rho\frac{d\mu'_k}{dt} - P_i\frac{dE'_i}{dt} - \frac{q_i}{\theta}\theta_{,i} + J'_iE'_i \geq 0.$$

For constitutive relations we break $\boldsymbol{\tau}$, $\mathbf{P}$ and $\mathbf{B}^L$ into recoverable and dissipative parts as [15]

$$\boldsymbol{\tau} = \boldsymbol{\tau}^R + \boldsymbol{\tau}^D, \quad \mathbf{P} = \mathbf{P}^R + \mathbf{P}^D, \quad \mathbf{B}^L = {}^R\mathbf{B}^L + {}^D\mathbf{B}^L.$$

(10.5.5)

The recoverable parts of Eq. (10.5.5) satisfy

$$\rho\frac{dF}{dt} = \tau_{ij}^R v_{j,i} - A_{ij}\rho\left(\frac{d\mu_j'}{dt}\right)_{,i} - {}^RB_k^L\rho\frac{d\mu_k'}{dt} - P_i^R\frac{dE_i'}{dt} - \rho\eta\frac{d\theta}{dt}. \quad (10.5.6)$$

Then the energy equation in Eq. (10.5.3) and the Clausius–Duhem inequality in Eq. (10.5.4) reduce to

$$\rho\theta\frac{d\eta}{dt} = \tau_{ij}^D v_{j,i} - {}^DB_k^L\rho\frac{d\mu_k'}{dt} - P_i^D\frac{dE_i'}{dt} + J_i'E_i' + \rho r - q_{i,i}, \quad (10.5.7)$$

$$\tau_{ij}^D v_{j,i} - {}^DB_k^L\rho\frac{d\mu_k'}{dt} - P_i^D\frac{dE_i'}{dt} - \frac{q_i}{\theta}\theta_{,i} + J_i'E' \geq 0. \quad (10.5.8)$$

Equation (10.5.7) is the dissipation equation. With

$$v_{j,i} = X_{M,i}\frac{d}{dt}(y_{j,M}), \quad \left(\frac{d\mu_j'}{dt}\right)_{,i} = X_{M,i}\frac{d}{dt}(\mu_{j,M}'), \quad (10.5.9)$$

we write Eq. (10.5.6) as

$$\rho\frac{dF}{dt} = \tau_{ij}^R X_{M,i}\frac{d}{dt}(y_{j,M}) - \rho^R B_j^L\frac{d\mu_j'}{dt}$$

$$- \rho A_{ij} X_{M,i}\frac{d}{dt}(\mu_{j,M}') - P_i^R\frac{dE_i'}{dt} - \rho\eta\frac{d\theta}{dt}. \quad (10.5.10)$$

Let

$$F = F(y_{j,M}; \mu_i'; \mu_{j,M}'; E_i'; \theta). \quad (10.5.11)$$

Then

$$\frac{dF}{dt} = \frac{\partial F}{\partial(y_{j,M})}\frac{d}{dt}(y_{j,M}) + \frac{\partial F}{\partial\mu_i'}\frac{d\mu_i'}{dt}$$

$$+ \frac{\partial F}{\partial(\mu_{j,M}')}\frac{d}{dt}(\mu_{j,M}') + \frac{\partial F}{\partial E_i'}\frac{dE_i'}{dt} + \frac{\partial F}{\partial\theta}\frac{d\theta}{dt}. \quad (10.5.12)$$

We substitute Eq. (10.5.12) into Eq. (10.5.10) and use Lagrange multipliers λ and L_M to introduce the constrains in Eqs. (10.1.16)$_{1,3}$. This yields

$$\rho\frac{\partial F}{\partial(y_{j,M})}\frac{d}{dt}(y_{j,M}) + \rho\frac{\partial F}{\partial\mu_i'}\frac{d\mu_i'}{dt}$$

$$+ \rho\frac{\partial F}{\partial(\mu_{j,M}')}\frac{d}{dt}(\mu_{j,M}') + \rho\frac{\partial F}{\partial E_i'}\frac{dE_i'}{dt} + \rho\frac{\partial F}{\partial\theta}\frac{d\theta}{dt}$$

$$= \tau_{ij}^R X_{M,i} \frac{d}{dt}(y_{j,M}) - {}^R B_j^L \rho \frac{d\mu_j'}{dt}$$

$$- \rho A_{ij} X_{M,i} \frac{d}{dt}(\mu_{j,M}') - P_i^R \frac{dE_i'}{dt} - \rho\eta \frac{d\theta}{dt}$$

$$+ \lambda \rho \mu_k' \frac{d\mu_k'}{dt} + L_M \rho \left(\mu_{k,M}' \frac{d\mu_k'}{dt} + \mu_k' \frac{d\mu_{k,M}'}{dt} \right), \quad (10.5.13)$$

or

$$\left[X_{M,i} \tau_{ij}^R - \rho \frac{\partial F}{\partial(y_{j,M})} \right] \frac{d}{dt}(y_{j,M})$$

$$- \left[P_i^R + \rho \frac{\partial F}{\partial E_i'} \right] \frac{dE_i'}{dt} - \rho \left[\eta + \frac{\partial F}{\partial\theta} \right] \frac{d\theta}{dt}$$

$$- \rho \left[{}^R B_i^L - \lambda \mu_i' - L_M \mu_{i,M}' + \frac{\partial F}{\partial\mu_i'} \right] \frac{d\mu_i'}{dt}$$

$$- \rho \left[X_{M,i} A_{ij} - L_M \mu_j' + \frac{\partial F}{\partial(\mu_{j,M}')} \right] \frac{d}{dt}(\mu_{j,M}') = 0. \quad (10.5.14)$$

Equation (10.5.14) implies the following recoverable constitutive relations:

$$X_{M,i} A_{ij} = -\frac{\partial F}{\partial(\mu_{j,M}')} + L_M \mu_j',$$

$$ {}^R B_i^L = -\frac{\partial F}{\partial\mu_i'} + \lambda \mu_i' + L_M \mu_{i,M}', \quad (10.5.15)$$

$$X_{M,i} \tau_{ij}^R = \rho \frac{\partial F}{\partial(y_{j,M})}, \quad P_i^R = -\rho \frac{\partial F}{\partial E_i'}, \quad \eta = -\frac{\partial F}{\partial\theta}.$$

Since we have assumed Eq. (10.1.13) which impose restrictions on **F** and $\mathbf{B}^L$, we can use Eq. (10.1.13) to determine L_M and λ [15]. From $(10.5.15)_1$ and Eq. (10.4.6) which is an implication of Eq. $(10.1.13)_1$, we have

$$X_{M,i} A_{ij} \mu_j' = -\frac{\partial F}{\partial(\mu_{j,M}')} \mu_j' + L_M \mu_j' \mu_j' = 0, \quad (10.5.16)$$

which determines

$$L_M = \frac{1}{\mu_s'^2} \frac{\partial F}{\partial(\mu_{k,M}')} \mu_k'. \quad (10.5.17)$$

Equation $(10.1.13)_2$ implies that

$$\mathbf{B}^L \cdot \mathbf{M}' = {}^R\mathbf{B}^L \cdot \mathbf{M}' + {}^D\mathbf{B}^L \cdot \mathbf{M}' = 0. \tag{10.5.18}$$

As a sufficient condition of Eq. (10.5.18), we let

$$^R\mathbf{B}^L \cdot \mathbf{M}' = 0, \quad {}^D\mathbf{B}^L \cdot \mathbf{M}' = 0. \tag{10.5.19}$$

From Eqs. $(10.5.19)_1$ and $(10.5.15)_2$,

$$^R B_i^L \mu_i' = -\frac{\partial F}{\partial \mu_i'}\mu_i' + \lambda \mu_i' \mu_i' + L_M \mu_i' \mu_{i,M}' = 0, \tag{10.5.20}$$

which determines

$$\lambda = \frac{1}{\mu_s'^2}\frac{\partial F}{\partial \mu_k'}\mu_k'. \tag{10.5.21}$$

With substitutions from Eqs. (10.5.17) and (10.5.21), the recoverable constitutive relations in Eqs. $(10.5.15)_{1,2}$ become

$$A_{ij} = -y_{i,M}\left[\frac{\partial F}{\partial(\mu_{j,M}')} - \frac{1}{\mu_s'^2}\frac{\partial F}{\partial(\mu_{k,M}')}\mu_k'\mu_j'\right],$$

$$^R B_i^L = -\frac{\partial F}{\partial \mu_i'} + \frac{1}{\mu_s'^2}\left[\frac{\partial F}{\partial \mu_k'}\mu_k'\mu_i' + \frac{\partial F}{\partial(\mu_{k,M}')}\mu_k'\mu_{i,M}'\right]. \tag{10.5.22}$$

To satisfy the rotational invariance (objectivity [1]) and Eq. (10.4.10), F can be reduced to a function of the following inner products [15]:

$$C_{KL} = y_{i,K}y_{i,L}, \quad G_{LM} = \mu_{i,K}'\mu_{i,M}',$$

$$N_L = y_{i,L}\mu_i', \quad W_L = y_{i,L}E_i'. \tag{10.5.23}$$

Instead of the deformation tensor C_{KL}, we use the strain tensor E_{KL}. Therefore, we take

$$F = F(E_{KL}; N_K; G_{LM}; W_K; \theta), \quad E_{KL} = \frac{1}{2}(C_{KL} - \delta_{KL}). \tag{10.5.24}$$

We assume that F is written symmetrically in the following sense:

$$\frac{\partial F}{\partial E_{KL}} = \frac{\partial F}{\partial E_{LK}}, \quad \frac{\partial F}{\partial G_{KL}} = \frac{\partial F}{\partial G_{LK}}. \tag{10.5.25}$$

In differentiating F, the elements of E_{KL} and G_{KL} are treated independently, i.e.,

$$\frac{\partial E_{KL}}{\partial E_{LK}} = 0, \quad \frac{\partial G_{KL}}{\partial G_{LK}} = 0, \quad K \neq L. \tag{10.5.26}$$

Then, it can be shown that

$$L_M = \frac{\partial F}{\partial(\mu'_{k,M})}\mu'_k = 0, \tag{10.5.27}$$

and the constitutive relations in Eqs. (10.5.22) and (10.5.15)$_{3-5}$ become

$$\tau_{ij}^R = \rho y_{i,M}\frac{\partial F}{\partial E_{ML}}y_{j,L} + \rho y_{i,M}\frac{\partial F}{\partial N_M}\mu'_j + \rho y_{i,M}\frac{\partial F}{\partial W_M}E'_j, \tag{10.5.28}$$

$$A_{ij} = -2y_{i,M}\mu'_{j,L}\frac{\partial F}{\partial G_{ML}}, \tag{10.5.29}$$

$$^R B_i^L = -y_{i,L}\frac{\partial F}{\partial N_L} + \frac{1}{\mu_s'^2}y_{k,L}\mu'_k\mu'_i\frac{\partial F}{\partial N_L}, \tag{10.5.30}$$

$$P_i^R = -\rho y_{i,L}\frac{\partial F}{\partial W_L}, \quad \eta = -\frac{\partial F}{\partial\theta}. \tag{10.5.31}$$

It can be verified that Eqs. (10.5.28)–(10.5.31) satisfy Eq. (10.4.19) (Noether's theorem).

10.6 Summary of Equations

In summary, the field equations are

$$\nabla \cdot \mathbf{D} = \sigma, \tag{10.6.1}$$

$$\nabla \cdot \mathbf{B} = 0, \tag{10.6.2}$$

$$\nabla \times \mathbf{E} = -\frac{\partial \mathbf{B}}{\partial t}, \tag{10.6.3}$$

$$\nabla \times \mathbf{H} = \mathbf{J} + \frac{\partial \mathbf{D}}{\partial t}, \tag{10.6.4}$$

$$\frac{1}{\gamma}\frac{d\mu'_i}{dt} = \varepsilon_{ijk}\mu'_j\left(-A_{lk,l} - \frac{A_{lk}}{\rho}\rho_{,l} + B_k + B_k^L\right), \tag{10.6.5}$$

$$\frac{d\rho}{dt} + \rho\nabla \cdot \mathbf{v} = 0, \tag{10.6.6}$$

$$\rho\frac{d\mathbf{v}}{dt} = \nabla \cdot \boldsymbol{\tau} + \rho\mathbf{f} + \mathbf{F}^{EM}. \tag{10.6.7}$$

The constitutive relations are determined by

$$\boldsymbol{\tau} = \boldsymbol{\tau}^R + \boldsymbol{\tau}^D, \quad \mathbf{P} = \mathbf{P}^R + \mathbf{P}^D, \quad \mathbf{B}^L = {}^R\mathbf{B}^L + {}^D\mathbf{B}^L, \quad (10.6.8)$$

$$F = F(E_{KL}; N_K; G_{LM}; W_K; \theta), \qquad (10.6.9)$$

and expressions for $\mathbf{q}$ and $\mathbf{J}'$. They are restricted by

$$^D\mathbf{B}^L \cdot \mathbf{M} = 0, \qquad (10.6.10)$$

$$\tau_{ij}^D v_{j,i} - {}^D B_k^L \rho \frac{d\mu_k'}{dt} - P_i^D \frac{dE_i'}{dt} - \frac{q_i}{\theta}\theta_{,i} + J_i'E' \geq 0. \qquad (10.6.11)$$

In addition, we have

$$\mathbf{D} = \varepsilon_0 \mathbf{E} + \mathbf{P}, \quad \mathbf{H} = \frac{\mathbf{B}}{\mu_0} - \mathbf{M}. \qquad (10.6.12)$$

On a boundary surface with an outward unit normal $\mathbf{n}$, possible boundary conditions are the prescriptions of [15]

$$\mathbf{y} \quad \text{or} \quad \mathbf{n} \cdot (\boldsymbol{\tau} + \mathbf{T}^{\text{EM}} + \mathbf{vG}),$$
$$\mathbf{n} \cdot \mathbf{D} \quad \text{and} \quad \mathbf{n} \times \mathbf{H}',$$
$$\mathbf{n} \cdot \mathbf{B} \quad \text{and} \quad \mathbf{n} \times \mathbf{E}', \qquad (10.6.13)$$
$$\theta \quad \text{or} \quad \mathbf{n} \cdot \mathbf{q},$$
$$\delta\theta \quad \text{or} \quad \mathbf{n} \cdot \mathbf{A} \times \boldsymbol{\mu}',$$

where

$$\mathbf{H}' \cong \mathbf{H} - \mathbf{v} \times \mathbf{D}. \qquad (10.6.14)$$

$\delta\theta$ is the angular displacement of $\boldsymbol{\mu}'$. In dynamic problems, instead of $\delta\theta$, the corresponding $\boldsymbol{\omega}$ may be prescribed.

Appendix 1

List of Symbols

X_K — reference or material coordinates

y_i — present or spatial coordinates

$\mathbf{i}, \mathbf{j}, \mathbf{k}$ $(\mathbf{I}, \mathbf{J}, \mathbf{K})$ — basis vectors of Cartesian coordinates

$\mathbf{e}_1, \mathbf{e}_2, \mathbf{e}_3$ — basis vectors of Cartesian coordinates

$\mathbf{e}_r, \mathbf{e}_\theta, \mathbf{e}_z$ — basis vectors of cylindrical coordinates

i, j, k, I, J, K — tensor indices. Range: 1–3

p, q — matrix indices. Range: 1–6

δ_{ij}, δ_{KL} — Kronecker delta

δ_{iK}, δ_{Ki} — coordinate transformation coefficients

$\varepsilon_{ijk}, \varepsilon_{IJK}$ — permutation symbol

$\mathbf{u}$ — displacement vector

J — Jacobian of deformation

C_{KL} — deformation tensor

E_{KL} — finite strain tensor

S_{kl} — linear strain tensor

v_i — velocity vector

a_i — acceleration vector

d_{ij} — deformation rate tensor

ω_{ij} — spin tensor

d/dt — material time derivative

ρ — present mass density

ρ^0 — reference mass density

τ_{kl} — Cauchy stress tensor

K_{Lj} — first Piola–Kirchhoff stress tensor

P_{KL} — second Piola–Kirchhoff stress tensor

T_{kl} — linear stress tensor

U — internal energy per unit volume

ε — internal energy per unit mass

U^F — electromagnetic field energy per unit volume

$\hat{U}$ — sum of field and internal energy densities $(= U^F + U)$

H — enthalpy per unit volume
χ — enthalpy per unit mass
F — free energy per unit mass
η — entropy per unit mass
θ — absolute temperature
q_i — heat flux vector
r — heat source per unit mass
φ — electrostatic potential
E_i — electric field
P_i — electric polarization per unit volume
D_i — electric displacement
ρ^t — total charge per unit volume
ρ^P — effective polarization charge per unit volume
σ^P — effective polarization charge per unit area
ρ^e — difference of total and polarization charges $(\rho^t - \rho^P)$
$p,\ n$ — concentrations of holes and electrons
$\gamma^p,\ \gamma^n$ — body sources of holes and electrons
$\mathbf{J}^p,\ \mathbf{J}^n$ — current densities of holes and electrons
B_i — magnetic flux or induction vector
H_i — magnetic field vector
M_i — magnetization per unit volume
μ_j — magnetization per unit mass
ψ — magnetostatic potential, angle of twist in torsion
$\mathbf{m}$ — magnetic moment
$\mathbf{f}^M$ — magnetic force
$\mathbf{T}^M$ — magnetic stress tensor
$\mathbf{c}^M$ — magnetic couple
w^M — magnetic body power per unit volume
S_i — poynting vector
$\mathbf{J}^t$ — total current density
$\mathbf{J}^M$ — effective magnetization current density
$\mathbf{J}$ — true current density $(\mathbf{J}^t - \mathbf{J}^M)$
$\mathbf{L}$ — angular momentum
$\mathbf{\Gamma}$ — couple or torque
$\mathbf{I}_G$ — centroidal moment of inertia tensor
ω_i — angular velocity of magnetization vector
ε_0 — vacuum electric permittivity
μ_0 — vacuum magnetic permeability
q — elementary charge

k_B — Boltzmann constant

γ — gyromagnetic ratio

c_{ijkl}, s_{ijkl} — elastic stiffness and compliance

χ_{ij}^e — electric susceptibility

ε_{ij} — dielectric constants

μ_{ij}^p, μ_{ij}^n — mobility of holes and electrons

D_{ij}^p, D_{ij}^n — diffusion constants of holes and electrons

M_s, M^0 — saturation magnetization per unit volume

μ_s, μ_s' — saturation magnetization per unit mass

χ_{ij}^M — magnetic susceptibility

μ_{ij} — magnetic permeability

$_2\chi_{KL}$ — second order magnetic constants

$_3\chi_{KLM}$ — third order magnetic constants

$_4\chi_{KLMN}$ — fourth order magnetic constants

h_{KIJ} — piezomagnetic constants

b_{KLMN} — magnetostrictive constants

γ_{KLMN} — exchangestrictive constants

f_{KLM} — magnetoexchange constants

α_{KL} — exchange constants

$\bar{c}_{ijkm}$ — effective elastic constants

$\bar{\chi}_{ik}$ — effective magnetic constants

$\bar{g}_{ikm}$, $\bar{h}_{kij}$ — effective piezomagnetic constants

$\bar{\beta}_{ijkl}$ — effective exchange constants

Appendix 2

SI and Gaussian Units

SI	Gaussian
ε_0	1
μ_0	1
$c = 1/\sqrt{\varepsilon_0\mu_0} = 2.9979 \times 10^8 \,\text{m/s}$	$c = 2.9979 \times 10^{10} \,\text{cm/s}$
$\mathbf{E} = \dfrac{Q\mathbf{r}}{4\pi\varepsilon_0 r^3}$	$\mathbf{E} = \dfrac{Q\mathbf{r}}{r^3}$
$\mathbf{B} = \displaystyle\int \dfrac{\mu_0 \mathbf{J}(\mathbf{x}') \times \mathbf{r}}{4\pi r^3} dV'$	$\mathbf{B} = \displaystyle\int \dfrac{\mathbf{J}(\mathbf{x}') \times \mathbf{r}}{cr^3} dV'$
$\mathbf{F} = Q(\mathbf{E} + \mathbf{v} \times \mathbf{B})$	$\mathbf{F} = Q\left(\mathbf{E} + \dfrac{\mathbf{v}}{c} \times \mathbf{B}\right)$
$\nabla \times \mathbf{E} = -\dfrac{\partial \mathbf{B}}{\partial t}$	$\nabla \times \mathbf{E} = -\dfrac{1}{c}\dfrac{\partial \mathbf{B}}{\partial t}$
$\nabla \times \mathbf{H} = \dfrac{\partial \mathbf{D}}{\partial t} + \mathbf{J}$	$\nabla \times \mathbf{H} = \dfrac{1}{c}\dfrac{\partial \mathbf{D}}{\partial t} + \dfrac{4\pi}{c}\mathbf{J}$
$\nabla \cdot \mathbf{D} = \rho^e$	$\nabla \cdot \mathbf{D} = 4\pi\rho^e$
$\nabla \cdot \mathbf{B} = 0$	$\nabla \cdot \mathbf{B} = 0$
$\mathbf{D} = \varepsilon_0\mathbf{E} + \mathbf{P}$	$\mathbf{D} = \mathbf{E} + 4\pi\mathbf{P}$
$\mathbf{B} = \mu_0\mathbf{H} + \mu_0\mathbf{M}$	$\mathbf{B} = \mathbf{H} + 4\pi\mathbf{M}$
$\mathbf{S} = \mathbf{E} \times \mathbf{H}$	$\mathbf{S} = \dfrac{c}{4\pi}\mathbf{E} \times \mathbf{H}$
$dw = \mathbf{E} \cdot d\mathbf{D} + \mathbf{H} \cdot d\mathbf{B}$	$dw = \dfrac{1}{4\pi}(\mathbf{E} \cdot d\mathbf{D} + \mathbf{H} \cdot d\mathbf{B})$

Appendix 3

Vector Identities

$$\mathbf{a} \times (\mathbf{b} \times \mathbf{c}) = (\mathbf{a} \cdot \mathbf{c})\mathbf{b} - (\mathbf{a} \cdot \mathbf{b})\mathbf{c}$$

$$(\mathbf{u} \times \mathbf{v}) \cdot (\mathbf{w} \times \mathbf{t}) = (\mathbf{u} \cdot \mathbf{w})(\mathbf{v} \cdot \mathbf{t}) - (\mathbf{u} \cdot \mathbf{t})(\mathbf{v} \cdot \mathbf{w})$$

$$\nabla(fg) = g\nabla f + f\nabla g$$

$$\nabla \cdot (f\mathbf{a}) = (\nabla f) \cdot \mathbf{a} + f(\nabla \cdot \mathbf{a})$$

$$\nabla \times (f\mathbf{a}) = (\nabla f) \times \mathbf{a} + f(\nabla \times \mathbf{a})$$

$$\nabla \cdot (\mathbf{a} \times \mathbf{b}) = (\nabla \times \mathbf{a}) \cdot \mathbf{b} - \mathbf{a} \cdot (\nabla \times \mathbf{b})$$

$$\nabla \times (\mathbf{a} \times \mathbf{b}) = (\mathbf{b} \cdot \nabla)\mathbf{a} - (\mathbf{a} \cdot \nabla)\mathbf{b} + (\nabla \cdot \mathbf{b})\mathbf{a} - (\nabla \cdot \mathbf{a})\mathbf{b}$$

$$\nabla \times (\nabla \times \mathbf{a}) = \nabla(\nabla \cdot \mathbf{a}) - \nabla^2 \mathbf{a}$$

For a closed curve C enclosing an area A with a unit normal $\mathbf{n}$:

$$\oint_C \mathbf{dl} \cdot \mathbf{G} = \int_A (\nabla \times \mathbf{G}) \cdot \mathbf{n} dA$$

$$\oint_C \mathbf{dl} \times \mathbf{G} = \int_A (\mathbf{n} \times \nabla) \times \mathbf{G} dA$$

For a closed surface S with a unit normal $\mathbf{n}$ enclosing a volume V:

$$\int_S \mathbf{n} \cdot \mathbf{G} dS = \int_V \nabla \cdot \mathbf{G} dV$$

Appendix 4

Material Constants

Vacuum electric permittivity $(\varepsilon_0) = 8.854 \times 10^{-12}\,\text{F/m}$

Vacuum magnetic permeability $(\mu_0) = 12.57 \times 10^{-7}\,\text{H/m}$

Elementary charge $(q) = 1.602 \times 10^{-19}\,\text{C}$

Boltzmann constant $(k_B) = 1.381 \times 10^{-23}\,\text{J/K}$

$k_B T/q = 0.0259\,\text{V}$ at room temperature $300\,\text{K}$ [50]

Planck's constant $(\hbar) = 1.05442 \times 10^{-27}\,\text{erg-sec}$

Aluminum nitride (AlN) [51]

$$\rho = 3260\,\text{kg/m}^3$$

$$[c_{pq}] = \begin{bmatrix} 345 & 125 & 120 & 0 & 0 & 0 \\ 125 & 345 & 120 & 0 & 0 & 0 \\ 120 & 120 & 395 & 0 & 0 & 0 \\ 0 & 0 & 0 & 118 & 0 & 0 \\ 0 & 0 & 0 & 0 & 118 & 0 \\ 0 & 0 & 0 & 0 & 0 & 110 \end{bmatrix} \times 10^9\,\text{N/m}^2$$

$$[e_{ip}] = \begin{bmatrix} 0 & 0 & 0 & 0 & -0.48 & 0 \\ 0 & 0 & 0 & -0.48 & 0 & 0 \\ -0.58 & -0.58 & 1.55 & 0 & 0 & 0 \end{bmatrix} \text{C/m}^2$$

$$[\varepsilon_{ij}] = \begin{bmatrix} 8.0 & 0 & 0 \\ 0 & 8.0 & 0 \\ 0 & 0 & 9.5 \end{bmatrix} \times 10^{-11}\,\text{F/m}$$

Barium titanate (BaTiO₃)[52]

$$\rho = 5800 \, \text{kg/m}^3$$

$$[c_{pq}] = \begin{bmatrix} 166.0 & 77.0 & 78.0 & 0 & 0 & 0 \\ 77.0 & 166.0 & 78.0 & 0 & 0 & 0 \\ 78.0 & 78.0 & 162.0 & 0 & 0 & 0 \\ 0 & 0 & 0 & 43.0 & 0 & 0 \\ 0 & 0 & 0 & 0 & 43.0 & 0 \\ 0 & 0 & 0 & 0 & 0 & 44.5 \end{bmatrix} \times 10^9 \, \text{N/m}^2$$

$$[e_{jq}] = \begin{bmatrix} 0 & 0 & 0 & 0 & 11.6 & 0 \\ 0 & 0 & 0 & 11.6 & 0 & 0 \\ -4.4 & -4.4 & 18.6 & 0 & 0 & 0 \end{bmatrix} \, \text{C/m}^2$$

$$[\varepsilon_{ij}] = \begin{bmatrix} 11.2 & 0 & 0 \\ 0 & 11.2 & 0 \\ 0 & 0 & 12.6 \end{bmatrix} \times 10^{-9} \, \text{F/m}$$

Cadmium selenide (CdSe) [53]

$$\rho = 4820 \, \text{kg/m}^3$$

$$[c_{pq}] = \begin{bmatrix} 90.7 & 58.1 & 51.0 & 0 & 0 & 0 \\ 58.1 & 90.7 & 51.0 & 0 & 0 & 0 \\ 51.0 & 51.0 & 93.8 & 0 & 0 & 0 \\ 0 & 0 & 0 & 15.04 & 0 & 0 \\ 0 & 0 & 0 & 0 & 15.04 & 0 \\ 0 & 0 & 0 & 0 & 0 & 16.3 \end{bmatrix} \times 10^9 \, \text{N/m}^2$$

$$[e_{ip}] = \begin{bmatrix} 0 & 0 & 0 & 0 & -0.21 & 0 \\ 0 & 0 & 0 & -0.21 & 0 & 0 \\ -0.24 & -0.24 & 0.44 & 0 & 0 & 0 \end{bmatrix} \, \text{C/m}^2$$

$$[\varepsilon_{ij}] = \begin{bmatrix} 9.02 & 0 & 0 \\ 0 & 9.02 & 0 \\ 0 & 0 & 9.53 \end{bmatrix} \varepsilon_0$$

Gallium arsenide (GaAs) [50, 53, 54]

$$\rho = 5307 \, \text{kg/m}^3$$

$$[c_{pq}] = \begin{bmatrix} 11.88 & 5.38 & 5.38 & 0 & 0 & 0 \\ 5.38 & 11.88 & 5.38 & 0 & 0 & 0 \\ 5.38 & 5.38 & 11.88 & 0 & 0 & 0 \\ 0 & 0 & 0 & 5.94 & 0 & 0 \\ 0 & 0 & 0 & 0 & 5.94 & 0 \\ 0 & 0 & 0 & 0 & 0 & 5.94 \end{bmatrix} \times 10^{10} \, \text{N/m}^2$$

$$[e_{ip}] = \begin{bmatrix} 0 & 0 & 0 & 0.154 & 0 & 0 \\ 0 & 0 & 0 & 0 & 0.154 & 0 \\ 0 & 0 & 0 & 0 & 0 & 0.154 \end{bmatrix} \, \text{C/m}^2$$

$$[\varepsilon_{ij}] = \begin{bmatrix} 12.5 & 0 & 0 \\ 0 & 12.5 & 0 \\ 0 & 0 & 12.5 \end{bmatrix} \times 8.85 \times 10^{-12} \, \text{F/m}$$

$$\mu^n = 8500 \, \text{cm}^2/\text{V} \cdot \text{s}$$
$$\mu^p = 400 \, \text{cm}^2/\text{V} \cdot \text{s}$$

Germanium (Ge) [53, 54]

$$\rho = 5327 \, \text{kg/m}^3$$

$$[c_{pq}] = \begin{bmatrix} 128.9 & 48.3 & 48.3 & 0 & 0 & 0 \\ 48.3 & 128.9 & 48.3 & 0 & 0 & 0 \\ 48.3 & 48.3 & 128.9 & 0 & 0 & 0 \\ 0 & 0 & 0 & 67.1 & 0 & 0 \\ 0 & 0 & 0 & 0 & 67.1 & 0 \\ 0 & 0 & 0 & 0 & 0 & 67.1 \end{bmatrix} \times 10^9 \, \text{N/m}^2$$

$$[\varepsilon_{ij}] = \begin{bmatrix} 0.1398932 & 0 & 0 \\ 0 & 0.1398932 & 0 \\ 0 & 0 & 0.1398932 \end{bmatrix} \times 10^{-9} \, \text{F/m}$$

$$\mu^n = 3900 \, \text{cm}^2/\text{V} \cdot \text{s}$$
$$\mu^p = 1900 \, \text{cm}^2/\text{V} \cdot \text{s}$$

Lithium niobate (LiNbO$_3$) [55, 56]

$$\rho = 4700\,\text{kg/m}^3$$

$$[c_{pq}] = \begin{bmatrix} 2.03 & 0.53 & 0.75 & 0.09 & 0 & 0 \\ 0.53 & 2.03 & 0.75 & -0.09 & 0 & 0 \\ 0.75 & 0.75 & 2.45 & 0 & 0 & 0 \\ 0.09 & -0.09 & 0 & 0.60 & 0 & 0 \\ 0 & 0 & 0 & 0 & 0.60 & 0.09 \\ 0 & 0 & 0 & 0 & 0.09 & 0.75 \end{bmatrix} \times 10^{11}\,\text{N/m}^2$$

$$[e_{ip}] = \begin{bmatrix} 0 & 0 & 0 & 0 & 3.70 & -2.50 \\ -2.50 & 2.50 & 0 & 3.70 & 0 & 0 \\ 0.20 & 0.20 & 1.30 & 0 & 0 & 0 \end{bmatrix}\,\text{C/m}^2$$

$$[\varepsilon_{ij}] = \begin{bmatrix} 38.9 & 0 & 0 \\ 0 & 38.9 & 0 \\ 0 & 0 & 25.7 \end{bmatrix} \times 10^{-11}\,\text{F/m}$$

Lithium tantalate (LiTaO$_3$) [55, 56]

$$\rho = 7450\,\text{kg/m}^3$$

$$[c_{pq}] = \begin{bmatrix} 2.33 & 0.47 & 0.80 & -0.11 & 0 & 0 \\ 0.47 & 2.33 & 0.80 & 0.11 & 0 & 0 \\ 0.80 & 0.80 & 2.75 & 0 & 0 & 0 \\ -0.11 & -0.11 & 0 & 0.94 & 0 & 0 \\ 0 & 0 & 0 & 0 & 0.94 & -0.11 \\ 0 & 0 & 0 & 0 & -0.11 & 0.93 \end{bmatrix} \times 10^{11}\,\text{N/m}^2$$

$$[e_{ip}] = \begin{bmatrix} 0 & 0 & 0 & 0 & 2.6 & -1.6 \\ -1.6 & 1.6 & 0 & 2.6 & 0 & 0 \\ 0 & 0 & 1.9 & 0 & 0 & 0 \end{bmatrix}\,\text{C/m}^2$$

$$\varepsilon_{ij}] = \begin{bmatrix} 36.3 & 0 & 0 \\ 0 & 36.3 & 0 \\ 0 & 0 & 38.2 \end{bmatrix} \times 10^{-11}\,\text{F/m}$$

PZT-2 [53]

$$\rho = 7600\,\mathrm{kg/m^3}$$

$$[c_{pq}] = \begin{bmatrix} 13.5 & 6.79 & 6.81 & 0 & 0 & 0 \\ 6.79 & 13.5 & 6.81 & 0 & 0 & 0 \\ 6.81 & 6.81 & 11.3 & 0 & 0 & 0 \\ 0 & 0 & 0 & 2.22 & 0 & 0 \\ 0 & 0 & 0 & 0 & 2.22 & 0 \\ 0 & 0 & 0 & 0 & 0 & 3.36 \end{bmatrix} \times 10^{10}\,\mathrm{N/m^2}$$

$$[e_{ip}] = \begin{bmatrix} 0 & 0 & 0 & 0 & 9.8 & 0 \\ 0 & 0 & 0 & 9.8 & 0 & 0 \\ -1.9 & -1.9 & 9.0 & 0 & 0 & 0 \end{bmatrix} \mathrm{C/m^2}$$

$$[\varepsilon_{ij}] = \begin{bmatrix} 504\varepsilon_0 & 0 & 0 \\ 0 & 504\varepsilon_0 & 0 \\ 0 & 0 & 260\varepsilon_0 \end{bmatrix}$$

PZT-4 [52]

$$\rho = 7600\,\mathrm{kg/m^3}$$

$$[c_{pq}] = \begin{bmatrix} 138.5 & 77.37 & 73.64 & 0 & 0 & 0 \\ 77.37 & 138.5 & 73.64 & 0 & 0 & 0 \\ 73.64 & 73.64 & 114.8 & 0 & 0 & 0 \\ 0 & 0 & 0 & 25.6 & 0 & 0 \\ 0 & 0 & 0 & 0 & 25.6 & 0 \\ 0 & 0 & 0 & 0 & 0 & 30.6 \end{bmatrix} \times 10^{9}\,\mathrm{N/m^2}$$

$$[e_{jq}] = \begin{bmatrix} 0 & 0 & 0 & 0 & 12.72 & 0 \\ 0 & 0 & 0 & 12.72 & 0 & 0 \\ -5.2 & -5.2 & 15.08 & 0 & 0 & 0 \end{bmatrix} \mathrm{C/m^2}$$

$$[\varepsilon_{ij}] = \begin{bmatrix} 13.06 & 0 & 0 \\ 0 & 13.06 & 0 \\ 0 & 0 & 11.15 \end{bmatrix} \times 10^{-9}\,\mathrm{F/m}$$

PZT-5A [52]

$$\rho = 7750 \, \text{kg/m}^3$$

$$[c_{pq}] = \begin{bmatrix} 99.201 & 54.016 & 50.778 & 0 & 0 & 0 \\ 54.016 & 99.201 & 50.778 & 0 & 0 & 0 \\ 50.788 & 50.788 & 86.856 & 0 & 0 & 0 \\ 0 & 0 & 0 & 21.1 & 0 & 0 \\ 0 & 0 & 0 & 0 & 21.1 & 0 \\ 0 & 0 & 0 & 0 & 0 & 22.6 \end{bmatrix} \times 10^9 \, \text{N/m}^2$$

$$[e_{jq}] = \begin{bmatrix} 0 & 0 & 0 & 0 & 12.322 & 0 \\ 0 & 0 & 0 & 12.322 & 0 & 0 \\ -7.209 & -7.209 & 15.118 & 0 & 0 & 0 \end{bmatrix} \text{C/m}^2$$

$$[\varepsilon_{ij}] = \begin{bmatrix} 15.3 & 0 & 0 \\ 0 & 15.3 & 0 \\ 0 & 0 & 15.0 \end{bmatrix} \times 10^{-9} \, \text{F/m}$$

PZT-5H [53]

$$\rho = 7500 \, \text{kg/m}^3$$

$$[c_{pq}] = \begin{bmatrix} 12.6 & 7.95 & 8.41 & 0 & 0 & 0 \\ 7.95 & 12.6 & 8.41 & 0 & 0 & 0 \\ 8.41 & 8.41 & 11.7 & 0 & 0 & 0 \\ 0 & 0 & 0 & 2.30 & 0 & 0 \\ 0 & 0 & 0 & 0 & 2.30 & 0 \\ 0 & 0 & 0 & 0 & 0 & 2.33 \end{bmatrix} \times 10^{10} \, \text{N/m}^2$$

$$[e_{jq}] = \begin{bmatrix} 0 & 0 & 0 & 0 & 17.0 & 0 \\ 0 & 0 & 0 & 17.0 & 0 & 0 \\ -6.5 & -6.5 & 23.3 & 0 & 0 & 0 \end{bmatrix} \text{C/m}^2$$

$$[\varepsilon_{ij}] = \begin{bmatrix} 1700\varepsilon_0 & 0 & 0 \\ 0 & 1700\varepsilon_0 & 0 \\ 0 & 0 & 1470\varepsilon_0 \end{bmatrix}$$

Silicon (Si) [53, 54]

$$\rho = 2332\,\text{kg/m}^3$$

$$[c_{pq}] = \begin{bmatrix} 16.57 & 6.39 & 6.39 & 0 & 0 & 0 \\ 6.39 & 16.57 & 6.39 & 0 & 0 & 0 \\ 6.39 & 6.39 & 16.57 & 0 & 0 & 0 \\ 0 & 0 & 0 & 7.956 & 0 & 0 \\ 0 & 0 & 0 & 0 & 7.956 & 0 \\ 0 & 0 & 0 & 0 & 0 & 7.956 \end{bmatrix} \times 10^{10}\,\text{N/m}^2$$

$$[\varepsilon_{ij}] = \begin{bmatrix} 11.7\varepsilon_0 & 0 & 0 \\ 0 & 11.7\varepsilon_0 & 0 \\ 0 & 0 & 11.7\varepsilon_0 \end{bmatrix}$$

$$\mu^n = 1500\,\text{cm}^2/\text{V}\cdot\text{s}$$
$$\mu^p = 450\,\text{cm}^2/\text{V}\cdot\text{s}$$

Yttrium iron garnet or YIG ($Y_3Fe_5O_{12}$) [18]

$$\rho = 5.172\,\text{g/cm}^3$$
$$c_{11} = 26.9 \times 10^{11}\,\text{dyn/cm}^2, \quad c_{12} = 10.77 \times 10^{11}\,\text{dyn/cm}^2$$
$$c_{44} = 7.64 \times 10^{11}\,\text{dyn/cm}^2,$$
$$b_{11} - b_{12} = 1.66 \times 10^2, \quad b_{44} = 1.66 \times 10^2$$
$$\chi = 3_4\chi_{12} - 4\chi_{11} = 3.36 \times 10^{-5}\,\text{Oe}^{-2}, \quad \alpha_{11} = 1.87 \times 10^{-11}\,\text{cm}^2$$
$$\gamma = -1.76 \times 10^7\,\text{Oe-cm}^2/\text{dyn-sec}, \quad M^0 = 1750/4\pi\,\text{G}$$

Zinc oxide (ZnO) [53, 54]

$$\rho = 5680\,\text{kg/m}^3$$

$$[c_{pq}] = \begin{bmatrix} 20.97 & 12.11 & 10.51 & 0 & 0 & 0 \\ 12.11 & 20.97 & 10.51 & 0 & 0 & 0 \\ 10.51 & 10.51 & 21.09 & 0 & 0 & 0 \\ 0 & 0 & 0 & 4.247 & 0 & 0 \\ 0 & 0 & 0 & 0 & 4.247 & 0 \\ 0 & 0 & 0 & 0 & 0 & 4.43 \end{bmatrix} \times 10^{10}\,\text{N/m}^2$$

$$[e_{ip}] = \begin{bmatrix} 0 & 0 & 0 & 0 & -0.48 & 0 \\ 0 & 0 & 0 & -0.48 & 0 & 0 \\ -0.573 & -0.573 & 1.32 & 0 & 0 & 0 \end{bmatrix} \text{C/m}^2$$

$$[\varepsilon_{ij}] = \begin{bmatrix} 8.55\varepsilon_0 & 0 & 0 \\ 0 & 8.55\varepsilon_0 & 0 \\ 0 & 0 & 10.2\varepsilon_0 \end{bmatrix}$$

$$\mu^n = 200 \, \text{cm}^2/\text{V} \cdot \text{s}$$
$$\mu^p = 180 \, \text{cm}^2/\text{V} \cdot \text{s}$$

References

[1] A. C. Eringen (1980). *Mechanics of Continua*. Robert E. Krieger Publishing Company. (Huntington, New York)

[2] I. S. Sokolnikoff (1964). *Tensor Analysis* (2nd ed.). John Wiley & Sons. (New York)

[3] H. F. Tiersten (1975). Nonlinear electroelastic equations cubic in the small field variables. *J. Acoust. Soc. Am.*, 57, 660–666.

[4] R. D. Mindlin (2006). *An Introduction to the Mathematical Theory of Vibrations of Elastic Plates* (J. S. Yang, Ed.). World Scientific Publishing Company. (Singapore)

[5] J. C. Baumhauer & H. F. Tiersten (1973). Nonlinear electroelastic equations for small fields superposed on a bias. *J. Acoust. Soc. Am.*, 54, 1017–1034.

[6] H. F. Tiersten (1995). On the accurate description of piezoelectric resonators subject to biasing deformations. *Int. J. Eng. Sci.*, 33, 2239–2259.

[7] H. F. Tiersten (1971). On the nonlinear equations of thermoelectroelasticity. *Int. J. Eng. Sci.*, 9, 587–604.

[8] A. T. Adams (1971). *Electromagnetics for Engineers*. Ronald Press Company. (New York)

[9] W. K. H. Panofsky & M. Phillips (1962). *Classical Electricity and Magnetism*. Addison-Wesley Publishing Company. (Reading, Massachusetts)

[10] J. D. Jackson (1990). *Classical Electrodynamics* (2nd ed.). John Wiley & Sons. (Singapore)

[11] L. D. Landau & E. M. Lifshitz (1984). *Electrodynamics of Continuous Media* (2nd ed.). Butterworth-Heinemann. (Linacre House, Jordan Hill, Oxford)

[12] H. F. Tiersten (1990). *A Development of the Equations of Electromagnetism in Material Continua*. Springer-Verlag. (New York)

[13] S. H. Guo (1979). *Electrodynamics*, Higher Education Press. (Beijing)

[14] G. E. Hay (1958). *Vector and Tensor Analysis*. Dover Publications. (New York)

[15] H. F. Tiersten & C. F. Tsai (1972). On the interaction of the electromagnetic field with heat conducting deformable insulators. *J. Math. Phys.*, 13, 361–378.

[16] H. F. Tiersten (1964). Coupled magnetomechanical equation for magnetically saturated insulators. *J. Math. Phys.*, 5, 1298–1318.

[17] H. F. Tiersten (1965). Variational principle for saturated magnetoelastic insulators. *J. Math. Phys.*, 6, 779–787.

[18] H. F. Tiersten (1965). Thickness vibrations of saturated magnetoelastic plates. *J. Appl. Phys.*, 36, 2250–2259.

[19] H. F. Tiersten (1969). Surface coupling in magnetoelastic interactions. In *Surface Mechanics* (pp. 120–142). ASME. (New York)

[20] D. F. Nelson (1979). *Electric, Optic, and Acoustic Interactions in Dielectrics.* John Wiley & Sons. (New York)

[21] M. Krawczyk, M. L. Sokolovskyy, J. W. Klos & S. Mamica (2012). On the formulation of the exchange field in the Landau-Lifshitz equation for spin-wave calculation in magnonic crystals. *Adv. Condens. Matter Phys.*, 764783.

[22] E. Y. Tsymbal. *Introduction to Solid State Physics.* https://unlcms.unl.edu/cas/physics/tsymbal/teaching/SSP-927/index.shtml

[23] J. S. Yang (2023). An alternative derivation of the Landau-Lifshitz-Gilbert equation for saturated ferromagnets, arXiv. https://arxiv.org/abs/2305.18232

[24] T. L. Gilbert (2004). A phenomenological theory of damping in ferromagnetic materials. *IEEE Trans. Magn.*, 40, 3443–3449.

[25] J. S. Yang (2023). A macroscopic theory of saturated ferromagnetic conductors, arXiv. https://arxiv.org/abs/2306.11525

[26] W. F. Brown Jr. (1965). Theory of magnetoelastic effects in ferromagnetism. *J. Appl. Phys.*, 36, 994–1000.

[27] W. F. Brown Jr. (1966). *Magnetoelastic Interactions.* Springer-Verlag. (New York)

[28] G. A. Maugin & A. C. Eringen (1972). Deformable magnetically saturated media. I. Field equations. *J. Math. Phys.*, 13, 143–155.

[29] G. A. Maugin & A. C. Eringen (1972). Deformable magnetically saturated media. II. Constitutive theory. *J. Math. Phys.*, 13, 1334–1347.

[30] S. M. Bakharev, M. A. Borich & S. P. Savchenko (2021). Caustic of magnetoelastic waves in elastically isotropic ferromagnets. *J. Magn. Magn. Mater.*, 530, 167862.

[31] Z. Gareyeva, R. Doroshenko & S. Seregin (2005). Resonances of standing magnetoelastic and elastic waves in ambilateral yttrium iron garnet film. In *Proceedings of the Third Moscow International Symposium on Magnetism* (pp. 36–39).

[32] T. Kobayashi, R. C. Barker, J. L. Bleustein & A. Yelon (1973). Ferromagnetoelastic resonance in thin films. I. Formal treatment. *J. Appl. Phys.*, 7, 3273–3285.

[33] T. Kobayashi, R. C. Barker & A. Yelon (1973). Ferromagnetoelastic resonance in thin films. II. Applications to nickel. *Phys. Rev. B*, 7, 3286–3297.

[34] J. S. Yang, X. S. Cao & W. H. Xu (2023). Bending of a saturated ferromagnetoelastic plate under a local mechanical load. *Acta Mech. Solida Sin.*, 36, 794–801.

[35] J. S. Yang (2010). *Antiplane Motions of Piezoceramics and Acoustic Wave Devices*. World Scientific Publishing Company. (Singapore)

[36] Q. G. Xia, J. K. Du & J. S. Yang (2024). Antiplane problems of saturated ferromagnetoelastic solids. *Acta Mech.*, 235, 533–541.

[37] Q. G. Xia, J. K. Du & J. S. Yang (2023). Propagation of coupled acoustic, electromagnetic and spin waves in saturated ferromagnetoelastic solids, arXiv. https://arxiv.org/abs/2307.09171

[38] J. S. Yang (2020). *Mechanics of Piezoelectric Structures* (2nd ed.). World Scientific Publishing Company. (Singapore)

[39] R. D. Mindlin (1961). High frequency vibrations of crystal plates. *Q. Appl. Math.*, 19, 51–61.

[40] H. F. Tiersten & R. D. Mindlin (1962). Forced vibrations of piezoelectric crystal plates. *Q. Appl. Math.*, 20, 107–119.

[41] R. D. Mindlin (1972). High frequency vibrations of piezoelectric crystal plates. *Int. J. Solids Struct.*, 8, 895–906.

[42] N. Li & J. S. Yang (2025). Coupling between elastic waves and magnetic spin waves in saturated ferromagnetoelastic plates. *Mech. Mater.*, 206, 105361.

[43] U. Hartmann (1999). Magnetic force microscopy. *Annu. Rev. Mater. Sci.*, 29, 53–87.

[44] D. Lee, K. W. Lee, J. V. Cady, P. Ovartchaiyapong & A. C. B. Jayich (2017). Topical review: spins and mechanics in diamond. *J. Opt.*, 19, 033001.

[45] T. Oeckinghaus, S. A. Momenzadeh, P. Scheiger, T. Shalomayeva, A. Finkler, D. Dasari, R. Stöhr & J. Wrachtrup (2020). Spin-phonon interfaces in coupled nanomechanical cantilevers. *Nano Lett.*, 20, 463–469.

[46] J. M. Gere & S. P. Timoshenko (1984). *Mechanics of Materials* (2nd ed.). Wadsworth Publishing Company. (Belmont, California)

[47] L. Meirovitch (1967). *Analytical Methods in Vibrations*. Macmillan Publishing Company. (New York)

[48] J. S. Yang (2023). *Mechanics of Functional Materials*. World Scientific Publishing Company. (Singapore)

[49] J. S. Yang (2023). A continuum theory of elastic-ferromagnetic conductors, arXiv. https://arxiv.org/abs/2307.16669

[50] R. F. Pierret (1996). *Semiconductor Device Fundamentals*. Pearson Education. (Uttar Pradesh, India)

[51] K. Tsubouchi, K. Sugai & N. Mikoshiba (1981). AlN material constants evaluation and SAW properties on AlN/Al$_2$O$_3$ and AlN/Si. In *Proceedings of IEEE Ultrasonics Symposium* (pp. 375–380).

[52] F. Ramirez, P. R. Heyliger & E. Pan (2006). Free vibration response of two-dimensional magneto-electro-elastic laminated plates. *J. Sound Vib.*, 292, 626–644.

[53] B. A. Auld (1973). *Acoustic Fields and Waves in Solids* (Vol. 1). John Wiley & Sons. (New York)

[54] S. M. Sze (1981). *Physics of Semiconductor Devices* (2nd ed.). John Wiley & Sons. (New York)

[55] A. W. Warner, M. Onoe & G. A. Couqin (1967). Determination of elastic and piezoelectric constants for crystals in class (3m). *J. Acoust. Soc. Am.*, 42, 1223–1231.

[56] H. F. Tiersten (1969). *Linear Piezoelectric Plate Vibrations*. Plenum Press. (New York)

Index

www.ingramcontent.com/pod-product-compliance
Lightning Source LLC
Chambersburg PA
CBHW050639190326
41458CB00008B/2344